••••••••••••••••••••••••••••

Blood Substitutes:
Principles, Methods, Products and Clinical Trials
Vol. 1

Blood Substitutes:
Principles, Methods, Products and Clinical Trials

Vol. 1

Author   *Thomas Ming Swi Chang,* Montreal, P.Q.

63 figures and 2 tables, 1997

Basel · Freiburg · Paris · London · New York ·
New Delhi · Bangkok · Singapore · Tokyo · Sydney

KARGER
LANDES
SYSTEMS

Thomas Ming Swi Chang, O.C., M.D., C.M., Ph.D., FRCP(C)
Director, Artificial Cells and Organs Research Centre
Professor of Physiology, Medicine and Biomedical Engineering
Career Investigator, Medical Research Council of Canada
Faculty of Medicine, McGill University, Montreal, P.Q., Canada
www.physio.mcgill.ca/artcell

Honorary President,
International Society for Artificial Cells, Blood Substitutes &
Immobilization Biotechnology

Editor-in-Chief, Artificial Cells, Blood Substitutes & Immobilization
Biotechnology, An International Journal

Library of Congress Cataloging-in-Publication Data
Chang, Thomas Ming Swi.
Blood substitutes : principles, methods, products, and clinical trials / author, Thomas Ming Swi Chang.
Tissue engineering
Includes bibliographical references and index.
1. Blood substitutes. I. Title.
[DNLM: 1. Blood Substitutes.   WH 450 C456b 1997]
RM171.7.C48 1997 615'.39 – dc21   DNLM/DLC
ISBN 3-8055-6584-4 (hard cover: alk. paper)

© Copyright 1997 by S. Karger AG, P.O. Box, CH–4009 Basel (Switzerland)
Printed in Switzerland on acid-free paper by Reinhardt Druck, Basel
ISBN 3-8055-6584-4

Contents

Preface

Concentrated effort to develop blood substitutes for public use was only seriously started after 1986 because of public concerns regarding HIV in donor blood. Most of the research papers and books related to blood substitutes have been published during this short time [1–469]. Thus, there is exciting progress ongoing in blood substitutes at present. Physicians, surgeons, specialists in transfusion medicine, biomedical researchers, administrators and other concerned publics are keen to follow up on progress in this area. Furthermore, many graduate students, research fellows and research scientists are starting to enter this area of research and development. Where can they find the most relevant material among the large number of publications? Nearly all the books in this area are multi–authored books written for specialists and researchers in the field and they cover details and opinions in highly specialized areas. As such, nonspecialists in the field are often confused by the different approaches, emphasis and opinions. Furthermore, for those starting to carry out research in the field, there is no book that describes those basic laboratory methods needed for starting research on blood substitutes. Furthermore, searching the literature is not easy since this is a very interdisciplinary area with publications spread over many disciplines. This book is written to fill this void and it is presented in two volumes.

The first volume contains a detailed overview and methods of blood substitutes

In order to have an uninterrupted overiview of the most relevant materials, the first volume is written by the same author. The author has prepared the first modified hemoglobin blood substitute 40 years ago and has continued research

in this and other areas of blood substitutes since that time. He has also been en-
couraging others to carry out research and develop blood substitutes. He is now
making use of his background in blood substitute to prepare the first volume.
The first volume contains the following materials:

1. Detailed discussions on the what, why, how and where of blood substitutes for the public, physicians, surgeons, specialists in blood transfusion, administrators, biomedical researchers, planners and organizers of health care. Top-
ics include answers to these questions: Why do we need blood substitutes? What
are the potential areas of clinical applications? What does it mean to blood supply and risk of infection? What are modified hemoglobin blood substitutes? What
properties of hemoglobin are important for modified hemoglobin? What are the
composition, functional and efficacy properties of modified hemoglobin? How
safe is modified hemoglobin? Which modified hemoglobins are being produced
industrially? What stage of clinical trial are they in now? What are
perfluorochemical–based blood substitutes? What are the next generations of
blood substitutes being studied now?

2. Specific examples (*given in italics*) of detailed principles and basic laboratory methods for those who want to start research in blood substitutes. Based
on his 30 years of training starting researchers in this area, the author has selected those materials that are most relevant. This include the structural–functional properties of modified hemoglobin. In addition, details are given on laboratory methods of preparation, analysis, efficacy studies, safety analysis, immunological studies, screening test and animal studies. These are basic procedures
that once mastered, will allow the starting researchers to extend and expand into
new areas of research in blood substitutes.

3. The 469 references at the end of the first volume should make it easier
for readers to look up details of their own interest.

The second volume contains two major types of material

1. In any rapidly advancing and highly interdisciplinary field, the problem
of a single-authored book is that no one can be an expert in every area. Experts
in selected fields are therefore invited to write specific chapters. Professor
Greenburg, a surgeon and researcher on blood substitutes will write a chapter on
the surgical indications of blood substitutes. Professor Reiss will expand on the
area of perfluorochemicals. Dr. Rudolph from the U.S. Naval Research Laboratory, who directs one of the two largest groups in the world on hemoglobin lipid
vesicles as blood substitute, will write a chapter on this topic. Dr. Alayash from
the FDA has contributed a chapter on his expertise, the toxicity aspects with emphasis on nitric oxide and oxygen radicals.

2. Many are interested in knowing what blood substitute products are in production or are being tested in clinical trials in patients for eventual use. These materials are not easily available in one reference source. This section contains all the blood substitutes that are now in Phase II or Phase III clinical trials in patients. Dr. S. Gould who has one of the longest continuous experiences in the development and clinical trial of blood substitutes, will describe the Northfield polyhemoglobin that is now in Phase III clinical trials. Dr. D.J. Nelson from Baxter will describe their diaspirin cross–linked hemoglobin that is now in Phase III clinical trials. The group from Somatogen will discuss their recombinant human hemoglobin that is now well into Phase II clinical trials. Dr. L.B. Pearce and Dr. M.S. Gawryl from Biopure have contributed a chapter on their bovine polyhemoglobin that is now well into phase II clinical trials. Drs. J.G. Adamson and C. Moore has written a chapter on Hemosol's o–raffinose polyhemoglobin that is now in Phase II clinical trials. In the area of perfluorochemicals, Dr. P. Keipert has put together a full description of Alliance's product that is now well into phase II clinical trials. This section therefore covers all the blood substitute products that are now in phase II and phase III clinical trials.

Acknowledgments

I would like to acknowledge the following: The Medical Research Council of Canada for the following awards: MRC Research Fellow; 1962–1965, MRC Scholar (career development award) 1965–1968; MRC Career Investigator (career award) 1968–1999, and research grants since 1965. The Quebec Ministry of Education and Science, for the Virage Award of Center of Excellence in Biotechnology since 1985. In addition, all the other agencies and groups, especially the Bayer/Canadian Red Cross Society Funds.

It has been a pleasure working with the many graduate students, fellows, research associates, technical staff and others throughout the many years. Those who have collaborated in the area of blood substitutes are listed as co–authors in the reference section of this book. I appreciate my son, Dr. Harvey Chang, an internist, for reading the manuscript to ensure that this book is also suitable for practicing physicians and surgeons and others who are not specialized researchers in blood substitutes.

It is a pleasure to acknowledge the excellent B.Sc. honors physiology program at McGill University in the 1950's. It emphasized interdisciplinary teaching of advanced chemistry and physiology. Stimulated by this, I started in 1956 to prepare artificial red blood cells on my own in my dormitory room. After obtaining promising results, I approached Sir Arnold Burgen. I must here thank him for helping to persuade the Physiology Chairman, Professor F.C. MacIntosh, to allow me to substitute this research for the regular honors B.Sc. Physiology research requirements. I must also acknowledge the chairman for letting me use the teaching laboratory to continue with this work during my medical school years. This is a good place to acknowledge the encouragement and support during this early period of Professor F.C. McIntosh, Professor A.V. Burgen and Professor S.G. Mason. Others who have encouraged me in this research in the earlier years included Professor D.V. Bates, Professor A. Burton, Professor O. Denstedt, Professor S. Freedman, Professor Leblond, Professor Kolff and many others.

Dedication

My wife, Lancy, for her selfless dedication, moral support and encouragement throughout my 40 years of research on blood substitutes and artificial cells.

My children, Harvey, Victor, Christine and Sandra.

My grandchildren.

Chapter 1

Blood Substitutes: Principles, Methods, Products and Clinical Trials,
by Thomas Ming Swi Chang. © 1997 Karger Landes Systems.

•••••••••••••••••••••••••••

Introduction

There are now two major groups of red blood cell substitutes: (1) modified hemoglobin and (2) perfluorochemicals.

Hemoglobin (Hb) is the major protein of red blood cells responsible for the transport of oxygen. However, hemoglobin molecules extracted from red blood cells cannot be used as blood substitutes. They break down in the body and they also cause toxicity. Biotechnological techniques of crosslinking, recombinant modification and microencapsulation of hemoglobin have resulted in blood substitutes that can replace red blood cells. Unlike red blood cells, pasteurization, ultrafiltration and chemical means can sterilize blood substitutes. This removes microorganisms responsible for AIDS, hepatitis, etc. Since they are free of RBC blood group antigens, there is no need for crossmatching or typing. This saves time and facilities and allows on–the–spot transfusion like the infusion of salt solution. Furthermore, they can be lyophilized and stored for a long time as a stable dried powder and reconstituted with salt solution just before use. Besides using hemoglobin (Hb) extracted from human RBC, other sources of Hb are available. These include bovine Hb, recombinant human Hb and transgenic human Hb.

Clinical trials (Phase I, II and III) in human on crosslinked hemoglobin and recombinant hemoglobin are ongoing. Potential applications include their use in cardiac, cancer, orthopedic and other surgery, traumatic injuries with severe hemorrhage from accidents, disasters or conflicts and other areas. With these first generation blood substitutes well into clinical trials in humans, researchers are now carrying out research into second and third generation modified hemoglobin blood substitutes.

Perfluorochemicals (PFC) are synthetic fluids in which oxygen can dissolve. They are made into fine emulsions for use as oxygen carriers. The biggest advantage is that it they are synthetic material that can be produced in large amounts. Furthermore, unlike biologicals, purity can be more easily controlled. On the other hand, unlike hemoglobin, they have a much lower capacity for carrying oxygen so the patient has to be breathing high oxygen. In modified hemoglobin, the patient only needs to breathe room air. The earlier problem of complement activation has now been resolved by substituting surfactant. Improved fluorochemicals (PFC) have also recently allowed a higher concentration of PFC to be used. Phase II clinical trials are being carried out especially to study whether they can be used to delay the need for blood transfusion in surgery especially when used with autologous blood techniques. One of the potential uses of PFC is for those patients whose religion does not allow them to use donor blood or product prepared from donor blood.

Chapter 2

Blood Substitutes: Principles, Methods, Products and Clinical Trials,
by Thomas Ming Swi Chang. © 1997 Karger Landes Systems.

•••••••••••••••••••••••••••
Why Do We Need Blood Substitutes?

Most readers would want an answer to the above question before spending time on other details related to blood substitutes. We now have first generation blood substitutes in the form of modified hemoglobin and fluorochemicals in clinical trials in patients. They are available in a sterilized form free from infectious organisms. These blood substitutes are free of blood group antigens and therefore do not need cross–matching or typing. This saves time and facilities and allows on–the–spot transfusion similar to the infusion of salt solution. Furthermore, modified Hb can be lyophilized and stored for a long time as a stable powder. It can be reconstituted with salt solution just before use. The presently available crosslinked hemoglobin works well when the patient is breathing room air.

Potential Areas of Clinical Applications

The first generation blood substitutes available now only persist in the circulation with a half time of up to 30 hours. This means that their present potential clinical uses are in five broad areas [31, 174, 180, 191, 249, 454, 471]: (1) Red blood cell replacement in surgical settings; (2) red blood cell replacement in emergencies; (3) shorter term red blood cell replacement in other conditions; (4) in other areas where they are more useful than red blood cells and (5) in patients whose religious beliefs do not allow them to use donor blood. These are summarized in the outline below.

(1) Surgery:
  Cardiopulmonary bypass surgery
  Trauma surgery
  Cancer surgery
  Orthopedic surgery
  Other elective surgery
(2) Emergency resuscitation of traumatic blood loss:
  Individual accidents like car accidents and others
  Disasters like earthquake, plane accidents and others
  Civilian and noncivilian casualties in major or minor conflicts
(3) Short term red blood cell replacement in other conditions
(4) Other medical applications requiring oxygen delivery:
  Cardioplegia
  Organ preservation
  Balloon angioplasties
  Thrombosis and embolisms
  Others
(5) Requirements for patients who cannot receive donor blood because of religious beliefs.

(1) Surgery

This is especially important in cardiopulmonary bypass surgery, trauma sur-gery, cancer surgery, orthopedic surgery and other elective surgery. For example, a patient's own blood (autologous blood) can be collected just before surgery and replaced with blood substitutes. During surgery, blood loss can be replaced by further blood substitutes. This can be very substantial in some types of sur-gery like trauma surgery where the blood replacement can be up to 100 units. At the completion of surgery, the autologous blood can be re–infused.

(2) Emergency resuscitation of traumatic blood loss

Examples include car and other accidents; disasters like earthquakes, plane accidents and others; civilian and noncivilian casualties in major or minor con-flicts. In these cases, blood substitutes can be used to replace blood loss in hem-orrhagic shock. This is especially important in situations when there are no time or facilities for cross–matching or lack of sufficient supply of donor blood for immediate use. This could be encountered in severe highway accidents in remote areas. This will be particularly the case in disasters like earthquakes and in civilian or noncivilian casualties in minor or major conflicts. With ongoing research to include enzymes in modified hemoglobin to prevent reperfusion injury, they may be less likely to cause reperfusion injuries than the use of red blood cells.

(3) Short term red blood cell support for other conditions

There are other conditions like chronic anemia due to various causes including hemolytic anemia and aplastic anemia. With increasing use of chemotherapy there is increasing frequency of bone marrow depression. For long–term replacement in these situations, donor blood would be needed. However, in some special cases blood substitutes may have some limited roles. One example is in severe hemolytic anemia. Another example is for temporary support when very rare blood groups are not immediately available. Another example is temporary support until bone marrow transplantation or recombinant human erythropoietin starts to take over. Another example is that modified hemoglobin stimulates erythropoiesis in the bone marrow. Furthermore, it is also a convenient source of iron.

(4) Medical applications where blood substitutes may be more effective than red blood cells

Blood substitutes, especially those in solution, would be superior to red blood cells in perfusion through obstructed vessels. Examples include myocardial infarction or stroke. In addition, with ongoing research to include higher concentrations of enzymes into modified hemoglobin, they may be less likely to cause reperfusion injuries than the use of red blood cells. The potential uses of blood substitutes in organ preservation and in those situations where hypothermia is used are other examples. The ability of hemoglobin in red blood cells to release oxygen at low temperature is severely affected. Some types of modified hemoglobin can be prepared to have acceptable P_{50} at low temperature.

(5) Patients whose religious beliefs do not allow them to use donor blood

Here blood substitutes are the only choice, particularly perfluorochemicals. Those modified hemoglobins prepared from red blood cells are likely not suitable. Recombinant hemoglobin from microorganisms may have potential.

Decreasing the Risks of Infection

In the U.S., whole blood and red blood cell transfusion now exceeds 12 million units [249]. This is comparable to the ratio for Canadian population [125]. With increasing precautions in blood collection, allogenic blood is now safer than ever [125, 380]. Unfortunately, since blood cannot be sterilized at present, transmission of infection is still possible at the following rates [125]:

> HIV 1/225,000 units
> Hepatitis B 1/200,000 units
> Hepatitis (others) 1/3,300 units
> HTLV I/II 1/50,000 units
> Other rare viral, parasitic and bacterial infections.

If we use the figure of 12 million units/year of blood transfused in the U.S.A., the above incidence translates into cases per year in the U.S.A. of HIV 50, hepatitis B 50, other hepatitis 4,000.

More recently, Schreiber et al [380] used another approach to analyze the risks of donor blood. He followed 586,507 persons (who provided more than 80% of the U.S. blood supply) who each donated blood more than once in the last two years. All the donors have passed all the screening tests used. Based on this they calculate the risk of infection due to the donation of blood during the infectious window period when the donors are still tested as negative by the screening tests. The risks are as follows:

> HIV 1/493,000
> Hepatitis B 1/63,000
> Hepatitis C 1/103,000
> HTLV 1/641,000

Thus, the risk of the transmission of the above viruses continues to be very low. It would become lower with increasing new screening tests. However, until blood can be sterilized, the public will still encounter a low risk of infection. Can we justify even this very low risk if we have blood substitutes that can be sterilized to remove all infective agents? Once a person becomes infected, it is no longer a matter of risk or statistics to this person.

Decreasing the Load on the Blood System

Transfusion Triggers

The public is still very much preoccupied with real and perceived risks of potential infections from blood transfusion. This has led to many changes in transfusion practice.

The minimal level of red blood cells in the patients before giving transfusion – "transfusion triggers" – is increasingly low. Sometimes this may impair tissue oxygenation in those patients with cardiovascular and pulmonary impairments [125]. Blood substitutes with no risks or no perceived risks might provide a wider margin of safety for these patients.

Predeposition of Autologous Blood

Before surgery, 4–6 units of blood are collected from the patient during the 4–6 weeks before elective surgery. This blood from the patient himself is available if needed during surgery. Autologous blood for elective surgery is favored. However, autologous blood collection is still less than 8% of all transfusions and is unlikely to exceed 10% of the allogeneic transfusion needs since this is only applicable to some situations [31, 174, 180, 191, 249, 452, 468]. Thus, even in combination with recombinant human erythropoietin, autologous blood is only indicated in some patients undergoing elective procedures. Several weeks are required to collect 4–6 units of blood. Major surgery like trauma, cardiovascular surgery, orthopedic procedures and others place a much larger demand on blood requirements of up to 100 units in just one severe surgical procedure for trauma [31, 174, 180, 191, 249, 452, 468].

Normovolemic Hemodilution and Intraoperative Autologous Blood

Another technique being used is normovolemic hemodilution. Blood is collected from the patient just before surgery. This is then replaced with a salt solution or plasma expander. The collected blood can then be infused during surgery if needed or reinfused after surgery. As discussed earlier, a blood substitute would be more effective than crystalloid or plasma expander. Another source of blood is intraoperative autologous blood (IAT). This is used in only a few surgical situations now.

However, despite these and other approaches, the need for allogeneic blood transfusion has not decreased. This could be related to increasing needs from an increasingly aging population that needs more surgical and medical treatment. Furthermore, increasingly aggressive cancer chemotherapy, transplantation and AIDS therapy requires more blood for transfusion. The margin between blood supply and demand had by 1989 become dangerously narrow [447]. Blood substitutes will help to decrease the load on donor blood supply.

Availability for Public Uses

The public can have blood substitutes available for: (1) red blood cell replacement in surgical settings; (2) red blood cell replacement in emergencies like severe hemorrhage in trauma or car accidents; (3) shorter term red blood cell replacement in other conditions; (4) in other areas where they are more useful than red blood cells; and (5) in patients whose religious beliefs do not allow them to use donor blood. These have been described in detail earlier.

Ready for Unexpected Demands on the Blood System

Enough donor blood has to be stored and made available for unforeseen situations. These range from mass traffic or airline accidents to major disasters like earthquakes, or civilian and noncivilian casualties resulting from minor or major conflicts and wars. Severe trauma resulting in bleeding and hemorrhagic shock is one of the problems related to the above situations. Unfortunately, much of the donor blood, if stored for these unexpected situations, would become outdated and need to be discarded.

Standard storage methods can only keep donor blood useful for about one month. Very expensive and laborious methods are required to freeze donor blood for long term storage. On the other hand, modified hemoglobin can be easily stored for years as lyophilized powder and reconstituted just before use. Thus blood substitutes could replace the need to store donor blood for such unforeseen situations and lessen the burden on donor blood supply. Furthermore, blood substitutes have other special advantages in these situations. This has already been discussed in the above section of clinical applications and in other reviews [31, 423, 452].

Potential Lowering of the Costs of the Blood System

It is expected that the costs of donor blood will increase very significantly [1, 25, 380]. In special applications, red blood cell units that require leucocyte reduction and γ irradiation further increase costs. Blood substitutes do not have leucocytes or other materials. Increased demands in donor selection, screening tests and regulatory requirements will continue to drive up the costs of donor blood collection. The potential costs of litigation and insurance add further expenses. On the other hand, the costs of blood substitutes will only decrease as industrial production becomes more routine.

Chapter 3

Blood Substitutes: Principles, Methods, Products and Clinical Trials,
by Thomas Ming Swi Chang. © 1997 Karger Landes Systems.

●●●●●●●●●●●●●●●●●●●●●●●●●●●
What Are Modified Hemoglobin Blood Substitutes?

Red blood cells are the best oxygen carrying material for transfusion. Hemoglobin is the oxygen carrying protein of red blood cells. Donor blood can become even more useful if hemoglobin can be extracted and used. This way, there is no need for crossmatching since there is no membrane blood group antigen. This allows for ease of transfusion especially in emergencies, major disasters or wars. There will also be no limit to storage time. Furthermore, extracted hemoglobin can be sterilized to remove infective microorganisms and viruses related to AIDS and hepatitis. Furthermore, other sources of hemoglobin may be possible. Unfortunately, we cannot use hemoglobin to substitute for red blood cells. Hemoglobin has been successfully crosslinked or microencapsulated to prevent these problems.

Why Can't Hemoglobin Itself be Used as a Blood Substitute?

Hemoglobin is the oxygen–carrying protein in the red blood cell. Amberson [8] lysed red blood cells to obtain hemoglobin for study in animals. The membrane stroma in lysed red blood cells is very toxic to the kidney. Rabiner extracted hemoglobin from red blood cells and removed cell membranes to form stroma–free hemoglobin [355]. This stroma–free hemoglobin (SF–Hb) was tested extensively in animal studies by DeVenuto's group [128], Moss' group [304] and other groups. Animal studies have shown that stroma–free hemoglobin is effective and apparently safe. Unfortunately, Savitsky's Phase I clinical trial in humans shows that SF–Hb has adverse effects on the renal and cardiovascular

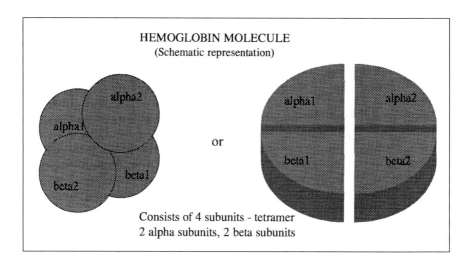

Fig. 1. Left: Perutz has defined the structure of hemoglobin shown schematically here. A hemoglobin molecule is a tetramer of four subunits: two α subunits and two β subunits. Right: author's simplified schematic representation for discussing modified hemoglobin. Reprinted with permission from Chang TMS. Biomaterials, Artificial Cells and Immobilization Biotechnology 1992; 20:154–174. Courtesy of Marcel Dekker Inc.

systems [379]. There are other reasons why hemoglobin is not suitable for use as a blood substitute. This is based on some of the basic properties of hemoglobin as discussed below.

Properties of Hemoglobin Inside and Outside the Red Blood Cells

The oxygen carrying protein of red blood cells, hemoglobin, has been extensively investigated by the pioneering work of Perutz [339, 340] and many others. Hemoglobin in the red blood cell is responsible for carrying and delivering oxygen to tissues. Perutz has carried out the most extensive studies on the structure–function properties of hemoglobin [339, 340]. A hemoglobin molecule is a tetramer of four subunits: two α (alpha) subunits and two β (beta) subunits (Fig. 1). Hemoglobin is in the "oxy," relaxed, or "R–state" when it carries oxygen. To release oxygen, the hemoglobin molecule undergoes conformational change with a 15° rotation. The molecule is then in the "deoxy," tense, or "T–

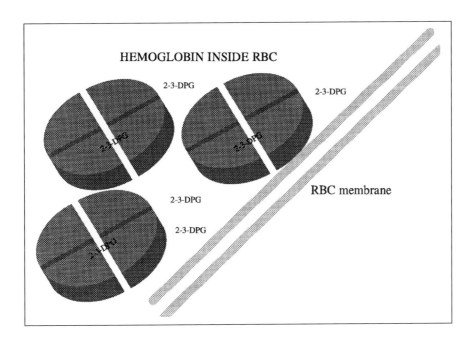

HEMOGLOBIN INSIDE RBC

2-3-DPG

2-3-DPG

2-3-DPG

2-3-DPG

2-3-DPG

2-3-DPG

2-3-DPG

2-3-DPG

RBC membrane

Fig. 2. Inside the red blood cell, hemoglobin stays as a tetramer. The red blood cell membrane retains 2,3–DPG in the cell to bind to hemoglobin. This allows hemoglobin to release oxygen as needed by the tissues with P_{50} of about 28 mmHg. Reprinted with permission from Chang TMS. Biomaterials, Artificial Cells and Immobilization Biotechnology 1992; 20:154–174. Courtesy of Marcel Dekker Inc.

state." The presence of the cofactor, 2,3–DPG, facilitates this conformational change. Thus, in the presence of the 2,3–DPG hemoglobin can release oxygen more readily at higher tissue oxygen tension (high P_{50}).

Hemoglobin Inside Red Blood Cells

Inside the red blood cell, hemoglobin is a tetramer. The red blood cell membrane retains 2,3–DPG in the cell to bind to hemoglobin (Fig. 2). This allows hemoglobin to more easily release oxygen. The hemoglobin concentration inside a red blood cell is 35 g/dl (14 g/dl in whole blood). This is possible because hemoglobin inside the cell does not exert an oncotic pressure in the plasma outside. Red blood cells have a lifetime in the circulation of about 100 days.

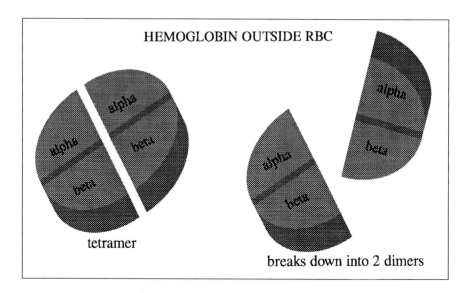

Fig. 3. When infused as free hemoglobin, each hemoglobin tetramer breaks down into two subunits – dimers. These smaller dimers are quickly removed from the circulation by the kidneys. Furthermore, the absence of 2,3–DPG outside the red blood cells means that hemoglobin cannot readily release oxygen (low P_{50}). Reprinted with permission from Chang TMS. Biomaterials, Artificial Cells and Immobilization Biotechnology 1992; 20:154–174. Courtesy of Marcel Dekker Inc.

Hemoglobin Outside Red Blood Cells (Including Stroma–Free Hemoglobin)

When hemoglobin is outside the red blood cells, the tetramer breaks down into two subunits – dimers (Fig. 3). These smaller dimers are filtered quickly through the kidney and removed from the circulation. These dimers are toxic to the kidney. There are also other problems when hemoglobin is outside the red blood cell. No 2,3–DPG is available in the plasma. Hemoglobin therefore cannot readily release oxygen until the tissue oxygen tension is much lower (low P_{50}). Furthermore, being free in the plasma, hemoglobin exerts an oncotic pressure. In order to have the same oncotic pressure as plasma, the concentration for hemoglobin solution has to be adjusted to 7 g/dl. This is compared to the whole blood hemoglobin concentration of 14 g/dl,

Modification of Hemoglobin by Microencapsulation and Crosslinking

The above problems with stroma–free hemoglobin mean that one has to modify it before use. The two major groups of modified hemoglobin are:

1. Microencapsulated hemoglobin
2. Crosslinked hemoglobin and recombinant hemoglobin.

Encapsulated Hemoglobin – Artificial Red Blood Cells

The first study on microencapsulated hemoglobin or artificial red blood cells was reported by Chang in 1957 [47]. In this approach, synthetic membranes are used to replace the biological membranes of red blood cells (Fig. 4). This way, hemoglobin inside the artificial red blood cells remains as a tetramer. The membrane used at this time was coated with a thin layer of organic liquid. This retained 2,3–DPG inside the artificial red blood cells. This allows them to take up and release oxygen with a P_{50} somewhat like that of hemoglobin in red blood (Fig. 5). Red blood cell enzymes like carbonic anhydrase [48] and catalase [54] when enclosed remained active. These artificial red blood cells do not have blood group antigens on the membrane. As a result, they do not form aggregates in the presence of blood group antibodies [59]. These early artificial red blood cells therefore fulfill most of the requirements of red blood cell substitutes.

However, there was one single major problem to be solved. This is the need to increase the circulation time after infusion into the animal. This has become the single major problem and earlier studies by Chang included the use of other synthetic polymers and crosslinked hemoglobin [48–52, 58–59]; membrane with surface charge [48–52, 58–59] and polysaccharide surface as sialic acid analogs [53, 56, 59]; lipid–protein and lipid–polymer [55, 59]. A major step forward in this regard was taken in 1980 when Djordjevich and Miller [528] reported their preparation of smaller submicron diameter lipid vesicles to encapsulate hemoglobin [59]. Many groups have since carried out research in the area of encapsulated hemoglobin as red blood cell substitutes using lipid membrane vesicles. These groups include Beissinger et al [18]; Domokos, Schmidt [133]; Farmer [146]; Gaber [160]; Hunt and Burnette [221, 212]; Miller et al [292]; Mobed and Chang [296–301]; Nishiya [323–325]; Phillips et al [342, 343]; Rudolph [101, 111, 372–374]; Szeboni et al [407–410]; Takahashi [412–414]; Tsuchida et al [428–431, 415, 416]; Usuuba [435–437]. A more recent approach using nanotechnology to prepare biodegradable hemoglobin nanocapsules was first reported in 1992 by Chang and Yu [87]. They have since reported further progress in this approach [102, 459–462]. Microencapsulated hemoglobin is

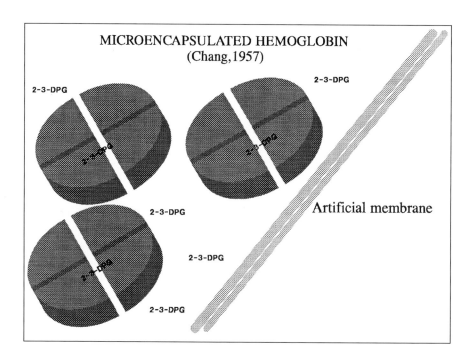

Fig. 4. The first study on microencapsulated hemoglobin or artificial red blood cells was reported by Chang in 1957 [47]. Here, synthetic membranes are used to replace the biological membranes of red blood cells. This allows the encapsulated hemoglobin to retain its tetrameric form and 2–3–DPG is also retained. Reprinted with permission from Chang TMS. Biomaterials, Artificial Cells and Immobilization Biotechnology 1992; 20:154–174. Courtesy of Marcel Dekker Inc.

considered the second and third generation of blood substitutes. Thus, this approach will be discussed in detail under a later chapter on the future development of blood substitutes.

In Chang's earlier studies to improve the circulation time of artificial red blood cells, one of his approaches is the use of crosslinked hemoglobin [48, 49]. This was used initially to form crosslinked hemoglobin membranes for artificial red blood cells. When he kept decreasing the diameter to about 1 micron, all the hemoglobin was crosslinked into polyhemoglobin [48, 49]. This will be discussed in the next section under crosslinked hemoglobin.

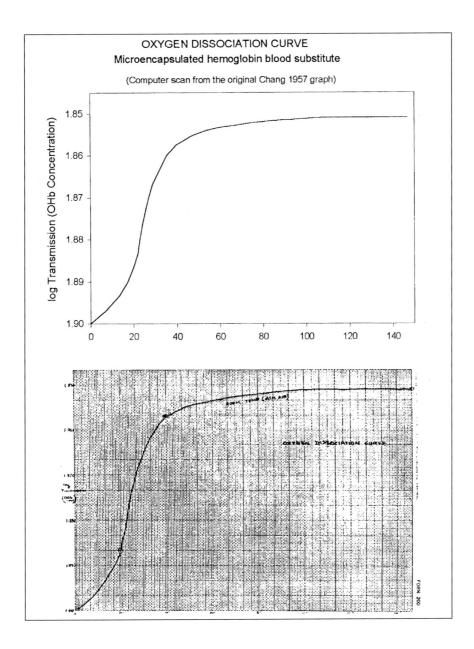

Fig. 5. Oxygen dissociation curve. Lower: original graph from this author's 1957 report [47]. – Upper: computer scan of original graph for printing. It shows the oxygen dissociation curve of the resulting artificial red blood cells. The 200 angstroms thin polymeric membrane was coated with a thin layer of organic liquid and this retained 2–3–DPG inside the artificial red blood cells. Reprinted with permission from Chang TMS, Report for Honors Physiology, McGill University Medical Library, 1957. Courtesy of author.

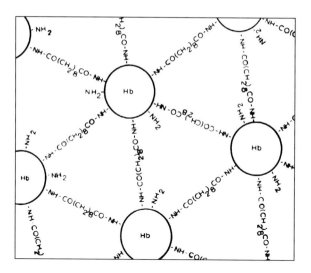

Fig. 6. In 1964 Chang reported the use of a bifunctional agent (sebacyl chloride) to crosslink hemoglobin to form crosslinked hemoglobin membranes and also polyhemoglobin [48]. As shown in this original 1965 figure by this author [49], the bifunctional agent crosslinked the many amino acids of the hemoglobin molecules. Reprinted with permission from Chang TMS, Ph.D. Thesis, McGill University, 1965. Also reprinted in Chang TMS, Artificial Cells, Charles C. Thomas Publisher, 1972 – with permission from copyright holder.

Crosslinked Hemoglobin

The three types of crosslinked hemoglobin are polyhemoglobin, conjugated hemoglobin and intramolecularly crosslinked hemoglobin.

Chang published the first report in 1964 on the use of a bifunctional agent, sebacyl chloride, to crosslink hemoglobin [48, 49]. This was used first to form crosslinked hemoglobin membranes for artificial red blood cells. He found that with decreasing size of artificial cells all the hemoglobin molecules are crosslinked into polyhemoglobin [48, 49]. At that time he proposed that the bifunctional agent crosslinked some of the many amino groups on the surface of the hemoglobin molecule (Fig. 6) [48, 49]. Crosslinking with a bifunctional agent prevents the breakdown of hemoglobin tetramers into dimers (Fig. 7). The reaction is as follows:

$$Cl-CO-(CH_2)_8-CO-Cl + \qquad HB-NH_2 = \qquad HB-NH-CO-(CH_2)_8-CO-NH-HB$$

Sebacyl Chloride $\qquad\qquad$ Hemoglobin $\qquad$ Crosslinked hemoglobin *(Chang 1964)*

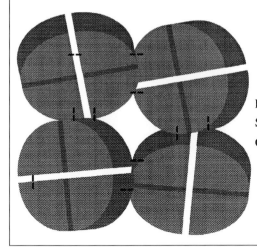

POLYHEMOGLOBIN: INTERMOLECULAR(+INTRA) XLINKAGE

First Xlinkers used:
Sebacyl chloride (Chang,1964,1965)
Glutaraldehye (Chang,1971)

Fig. 7. Crosslinking the hemoglobin molecules together presents their breakdown into dimers. Following the use of sebacyl chloride [48], the same author reported in 1971 the use of glutaraldehyde as another crosslinker [57]. Since then, many other investigators have extended and improve this using other crosslinkers. Reprinted with permission from Chang TMS. Biomaterials, Artificial Cells and Immobilization Biotechnology 1992; 20:154–174. Courtesy of Marcel Dekker Inc.

In 1971 Chang reported the use of another bifunctional agent, glutaraldehyde, to crosslink hemoglobin and a red blood cell enzyme, catalase [57]. The reaction is as follows:

$$H–CO–(CH_2)_3–CO–H + \quad HB–NH_2 = \quad HB–NH–CO–(CH_2)_3–CO–NH–HB$$
Glutaraldehyde   Hemoglobin   Crosslinked hemoglobin *(Chang 1971)*

In this 1971 procedure, catalase was added to 10 g/dl of hemoglobin and enclosed in 1.5 ml of microcapsules. To this was added 100 ml of a solution of 0.1 mol/L metaborate and 0.001 mol/L benzamideine HCl. Then, 200 µl of 50% glutaraldehyde (5.56 mol/L) was added and the mixture was kept slightly agitated at 20°C for 1 hour. Then 100 ml of 0.05 mol/L sodium borohydride was added and left at 4°C for 20 minutes, then washed twice with 200 ml of 0.154 mol/L sodium chloride solution. Crosslinking hemoglobin and catalase with glutaraldehyde this way increases the stability of the enzymes [57]. In this approach crosslinking was adjusted so that the crosslinked proteins are in a soluble state. This results in less steric hindrance and greater ease of substrate diffusion. This

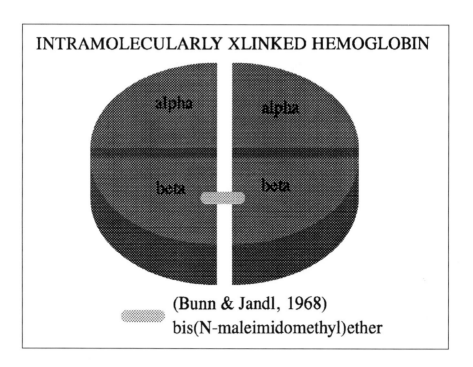

Fig. 8. Bunn and Jandl in 1968 used a bifunctional crosslinker, bis(N–maleim-idomethyl)ether, to crosslink inside each hemoglobin molecule between the β subunits. The resulting intramolecularly crosslinked hemoglobin tetramer is stable and does not breakdown into dimers. This has been extended by others including the use of diaspirin to crosslink the α subunits. Reprinted with permission from Chang TMS. Biomaterials, Artificial Cells and Immobilization Biotechnology 1992; 20:154–174. Courtesy of Marcel Dekker Inc.

procedure is the first reported study of the use of glutaraldehyde to crosslink protein molecules to one another (intermolecular). It can also crosslink the protein internally (intramolecular). In 1973 Payne used an extension of this glutaraldehyde method to crosslink protein to form soluble molecular weight markers for separation studies [337].

Bunn and Jandl in 1968 reported the use of a bifunctional crosslinker, bis(N–maleimidomethyl)ether, to crosslink the hemoglobin molecules internally (intramolecular) (Fig. 8) [38]. This prevents the breakdown of the tetramer into dimers and allows it to circulate longer so it is not excreted by the kidney.

Chang in 1964 reported crosslinking hemoglobin to polymers to form insoluble conjugated hemoglobin (Fig. 9, 10) [48, 49]. This has been extended to soluble conjugated hemoglobin (Fig. 10) formed by linking one hemoglobin

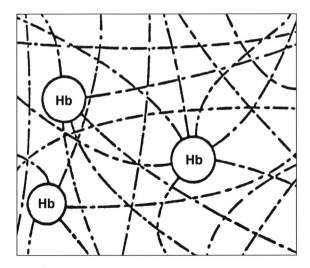

Fig. 9. Hemoglobin can also be conjugated to polymer to form artificial cell membrane or conjugated hemoglobin, as first reported by Chang in 1964 [48]. This is from the original 1965 figure [49, 59]. Reprinted with permission from Chang TMS.

molecule to soluble polymers as prepared by Sunder in 1975; Wong in 1976 [417, 462]; Iwashita in 1980 [220, 221]; the Enzon group in 1988 [397, 398].

Benesch et al in 1975 [20] made the important finding that the addition of a 2,3–DPG analog, pyridoxal phosphate, to hemoglobin can also improve the P_{50}. This is therefore added to the different types of crosslinked hemoglobin (Fig. 11).

As seen from Figure 12, most of the basic principles for modified hemoglobin were already available by the early 1970s. Unfortunately there was no public demand for or interest in these approaches then. The major efforts at that time were on stroma–free hemoglobin and perfluorochemicals. Savitsky et al's study in humans in 1978 showed that stroma–free hemoglobin was unsuitable for use in humans. Shortly after this, it was found that perfluorocarbons had some adverse effects in humans and it did not carry enough oxygen for many of the clinical applications. Thus, a few investigators started to turn to the earlier approach of encapsulated and crosslinked hemoglobin. This effort was suddenly intensified after 1986, when HIV in donor blood became a major public concern. As a result, extensive studies of modified hemoglobin have been carried out by many groups (Fig. 13). Modern development of crosslinked hemoglobin is based on an understanding of the structure–function relationship of hemoglobin. As a result, it is important to first discuss how this relates to modified hemoglobin.

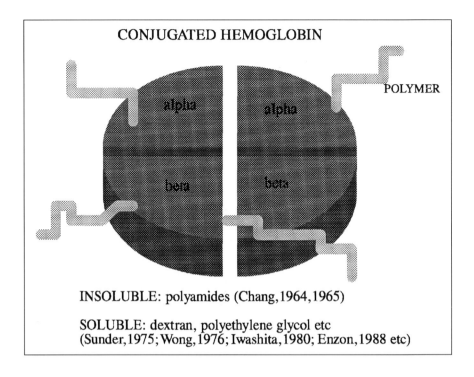

Fig. 10. Conjugated hemoglobin prevents the breakdown of tetramer into dimers. The surrounding polymer also masks the antigenic sites of the hemoglobin. This has been extended to soluble conjugated hemoglobin formed by linking one hemoglobin molecule to soluble polymers by Sunder, Wong, Iwashita and the Enzon group. Reprinted with permission from Chang TMS. Biomaterials, Artificial Cells and Immobilization Biotechnology 1992; 20:154–174. Courtesy of Marcel Dekker Inc.

Importance of Structure–Function Properties of Hemoglobin in Modified Hemoglobin

The lack of extensive research activities on crosslinked and encapsulated Hb between 1970 and 1986 had its advantages. Before 1965, most of the work on modified hemoglobin was done empirically without the availability of important basic knowledge in the structure-function relationship of the hemoglobin molecule. By 1980, Perutz [339, 240] and many others contributed extensively to the basic knowledge of the structure and function of hemoglobin. This allowed scientists to come up with novel approaches to crosslinking and modifying the

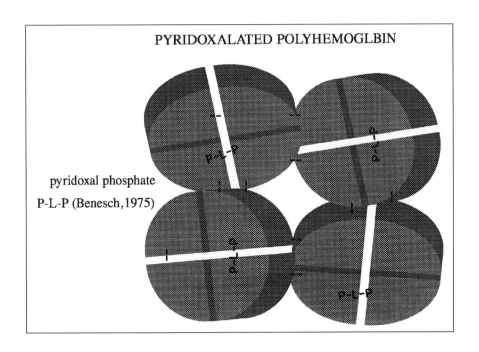

Fig. 11. Benesch's group in 1975 made the important finding that the addition of pyridoxal phosphate to hemoglobin can fulfill the function of 2,3–DPG. This made possible the improved usefulness of crosslinked hemoglobin. Reprinted with permission from Chang TMS. Biomaterials, Artificial Cells and Immobilization Biotechnology 1992; 20:154–174. Courtesy of Marcel Dekker Inc.

MODIFIED HEMOGLOBIN
(Ideas first reported)

1957(Chang)   ENCAPSULATED HB

1964(Chang)   CROSSLINKED POLYHB

1968(Bund)   TETRAMERIC XLINKED HB

1975(Benesch) PYRIDOXALATION

Not much interest in these at that time,
since empahsis was on stroma-free Hb & fluorocarbons

Fig. 12. The basic ideas for modified hemoglobin were already available many years ago. Lack of interest in this has prevented its development.

SFH 1978, HIV 1986: **STIMULATED RESEARCH INTO MODIFIED HB** **CROSSLINKED HEMOGLOBIN** **Agishi, Alayash, Bakker, Benesch, Biro, Bonhard, Bucci** **Chang, DeVenuto, DeWoskin, Estep, Faivre, Feola,** **Gould, Greenburg, Hess, Hori, Hsia, Hughes, Iwashita,** **Jacobs, Jesch, Keipert, Klugger, Magnin, Manning, McKenzie,** **Messmer, Moss, Nose, Przybelski, Pluira, Sekiguchi,** **Sideman, Shorr, Valeri, Winslow, Wong, Yang and others** **RECOMBINANT AND TRANSGENIC HEMOGLOBIN** **Caspari, Hoffman, Kumar, Looker, Olson, Shih,** **Shoemaker, Stetler and others** **ENCAPSULATED HEMOGLOBIN** **Beissinger, Chang, Djordjevich, Farmer, Feuersteom,** **Hunt, Kondo, Kobayashi, Lee, Ogata, Phillips, Rabinovic, Rudolph,** **Schmidt, Snohara, Szebeni, Takahasi,** **Takaori, Tsuchida, Usuba, Yu and others**

Fig. 13. After stroma–free hemoglobin was shown to be unsafe in humans in 1978, there were a few groups working on modified hemoglobin. However, it was public concern of HIV in donor blood in 1986 that stimulated many groups to develop this in the last 10 years.

hemoglobin molecule. The following section is a brief overview of the most relevant aspects of the structural–functional properties of hemoglobin as related to modified hemoglobin.

Reactive Amino Groups

Hemoglobin contains many reactive amino groups that can form the basis for crosslinking and other modifications (Fig. 6). Besides those in the 2,3–DPG pocket, there are 40 reactive lysines and 4 other reactive amino groups in each hemoglobin molecule. Many of the lysine groups are on the surface of the hemoglobin molecules. These reactive groups on the surface take part in the intermo-

lecular crosslinking reactions described earlier. Crosslinking or modification of specific reactive amino groups can result in modified hemoglobin with specific properties. This forms the basis of different chemical and recombinant techniques to modify hemoglobin. Possible variations are almost limitless, especially in view of the variety of crosslinking methods. Considering that hemoglobin S in sickle–cell anemia differs from the normal hemoglobin A in only one amino acid, it is obvious what significant changes can result from the different possible modifications. This is especially important when we look at the following structure–function properties of hemoglobin.

Subunits of a Hemoglobin Molecule

Each hemoglobin molecule consists of four polypeptide chains (Fig. 1). Hemoglobin A, the principal hemoglobin in adults, contains two α polypeptide chains and two β polypeptide chains. Each polypeptide chain contains a heme group and a single oxygen–binding site. These four chains are held together by noncovalent attractions. The hemoglobin molecule is nearly spherical having a diameter of 55 angstroms. Each α chain is in contact with both β chains. However, few interactions are present between the two α chains or between the two β chains.

Oxygen Affinity and 2,3–DPG

Oxygen affinity of hemoglobin is characterized by P_{50}. P_{50} is the partial pressure of oxygen at which hemoglobin is 50% saturated with O_2. P_{50} measures the ease of hemoglobin in unloading oxygen as required in the capillaries supplying the tissues. The higher the P_{50} the more ease it has in unloading oxygen. Hemoglobin in red blood cells has a P_{50} of 26 mmHg (26 Torr). An oxygen dissociation curve is useful in measuring many characteristics of hemoglobin (Fig. 14).

The affinity of hemoglobin for O_2 is regulated by organic phosphate such as 2,3–diphosphoglyerate (2,3–DPG). Benesch and Benesch showed in 1967 that in the presence of 2,3–DPG, hemoglobin has a higher P_{50}. This highly anionic organic phosphate is present in human red cells at the same molar concentration as hemoglobin. As a result, the P_{50} of hemoglobin inside red blood cells, 26 mmHg, is much higher than that of free hemoglobin outside the red blood cells. Hemoglobin in the red blood cells can therefore readily unload oxygen in capillaries supplying tissues.

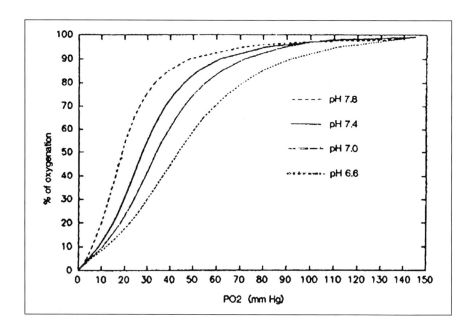

Fig. 14. This is a typical oxygen dissociation curve. Hemoglobin is 100% saturated with oxygen at an oxygen tension of 100 mmHg in the lung capillaries. Hemoglobin releases oxygen when it reaches the tissue capillaries where the oxygen tension is lower. P_{50} is the oxygen tension at which the hemoglobin releases 50% of its oxygen. The sigmoidal curve is due to the cooperativity of hemoglobin. The Bohr effect is shown here as the shifting of the curve to the right as pH decreases. Reprinted with permission from Yu WP, Chang TMS. Artificial Cells, Blood Substitutes and Immobilization Biotechnology, an International Journal, 1996; 24:169–184. Courtesy of Marcel Dekker Publisher.

The binding site for 2,3–DPG is important because it forms the basis of many modifications of hemoglobin to form modified hemoglobin blood substitutes. The 2,3–DPG pocket is in the central cavity of deoxyhemoglobin. The binding site for 2,3–DPG consists of three positively charged residues on each β chain: the α–amino group, lysine EF6, and histidine H21. These positively charged groups in the central cavity of the hemoglobin molecule interact with the strongly negatively charged 2,3–DPG. In this way, 2,3–DPG stabilizes the deoxyhemoglobin quaternary structure by crosslinking the β chains. This shifts the equilibrium toward the T (tense) form. In modified hemoglobin many approaches are available to optimize the P_{50} by modification of the 2,3–DPG pocket.

Cooperativity and Hill Coefficient

Oxygen binds cooperatively to hemoglobin because the binding of O_2 to a hemoglobin molecule enhances the binding of further O_2 to the same molecule. In other words, binding at one heme of one subunit facilitates the binding of oxygen at the other hemes of the other subunits in the same hemoglobin molecule. The reason for this is that when oxygen binds to a subunit, there is a structural change within this subunit. This is translated into structural changes at the interfaces between subunits. Thus, there is communication between the heme groups of a hemoglobin molecule – heme–heme interaction. Cooperativity allows each hemoglobin molecule to become a more efficient O_2 carrier that can deliver 1.83 times more oxygen. This cooperativity is shown in the sigmoidal shape of the oxygen dissociation curve of hemoglobin (Fig. 14).

The quantitative measurement of cooperativity is the Hill coefficient, n. n increases with an increase in cooperativity. The maximum value of n is the number of binding sites for oxygen in a hemoglobin molecule, total of 4. A Hill coefficient of 1 means that there is no cooperativity. For hemoglobin, the Hill coefficient is normally 2.8. This shows that hemoglobin binds oxygen cooperatively. When the Hill coefficient of hemoglobin is below 2.8, cooperativity is less than normal. In the modification of hemoglobin, the best configuration is to fix hemoglobin in the deoxyhemoglobin state (T or tense–state). In the T–state, the four subunits are most stable and P_{50} is also higher. However, this has to be properly adjusted in order not to restrict the structural changes at the interfaces between the subunits. Too much restriction will result in a decrease in cooperativity.

The Bohr Effect

The affinity of hemoglobin for O_2 is modified by pH and CO_2. Under physiological conditions, lowering the pH increases P_{50} and the oxygen dissociation curve is shifted to the right (Fig. 14). Increasing CO_2 without changing pH, also increases P_{50} and shifts the oxygen dissociation curve to the right. This property of right shifting of hemoglobin due to lower pH or increasing CO_2 is called the Bohr effect. This has important physiological significance. Active tissues such as contracting muscle, release more CO_2 and acid, resulting in an increase of these in the surrounding environment. The Bohr effect allows hemoglobin to unload more oxygen required by the active tissues.

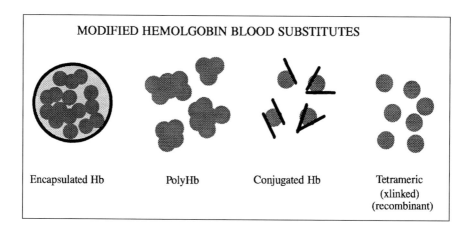

MODIFIED HEMOLGOBIN BLOOD SUBSTITUTES

Encapsulated Hb   PolyHb   Conjugated Hb   Tetrameric
(xlinked)
(recombinant)

Fig. 15. Summary of the four different types of modified hemoglobin: (1) encapsulated, (2) polyhemoglobin, (3) conjugated hemoglobin and (4) tetrameric hemoglobin – either intramolecularly crosslinked or intramolecularly modified by recombinant technology. Reprinted with permission from Chang TMS. Artificial Cells, Blood Substitutes and Immobilization Biotechnology, an International Journal 1997; 25:1–24. Courtesy of Marcel Dekker Inc.

Principles and Types of Crosslinked Hemoglobin

Figure 15 summarizes the four major groups of modified hemoglobin. Encapsulated hemoglobin is a second generation modified hemoglobin. It is therefore discussed in a later chapter under future perspectives. In the case of crosslinked hemoglobin, the basic chemistry of crosslinking protein on their reactive amino groups is not new. Its application in medicine has been described in details in two books as far back as 1976 [63]. Only those that are being developed for clinical trials are described below. More details are available in later chapters.

Glutaraldehyde as Crosslinker for Pyridoxalated Polyhemoglobin

The 1971 glutaraldehyde approach of Chang [57] has been developed as pyridoxalated polyhemoglobin blood substitutes by Dudziak and Bonhard in 1976 [134]; Moss, Gould and Sehgal in 1980 [383]; DeVenuto and Zegna in 1982 [129]; Chang's group in 1982 [234]; Stabilini et al in 1983 [405]; Hobbhahn et al in 1985 [194]. This type of polyhemoglobin requires the use of Benesch's pyridoxal phosphate to substitute for the 2,3–DPG [20]. The group of Gould and

Moss [383–386, 171–176, 304–305] has developed this very extensively to its present refined state. At present, the product from this group is effective when infused in a 3000 ml volume into patients in phase II clinical trials [176]. They are now in phase III clinical trials. This will be discussed in a later section and in a chapter in the second volume by Gould's group. This approach has also been developed extensively by the Biopure group using bovine hemoglobin to form polyhemoglobin now in phase II clinical trials [206–210, 222]. This will also be discussed later and in a separate chapter by this group.

Other Crosslinking Agents

As discussed earlier, basic research has resulted in increasing understanding of the structure–function relationship of hemoglobin. This has made it possible for many groups to carry out excellent in–depth studies to design specific crosslinkers and modifiers (Fig. 16). This has resulted in exciting improvements in the specificity of crosslinkers that function to modify the hemoglobin to improve its oxygen releasing properties. Those developed to a stage for clinical trials include the following.

Walder et al in 1979 reported the use of a 2,3–DPG pocket modifier, bis (3,5–dibromosalicyl) fumarate (DBBF) to intramolecularly crosslink the two α subunits of the hemoglobin molecule [446]. This prevents dimer formation and improves P_{50}. Baxter has extensively developed this [39, 141, 147] to the present product that is now in phase III clinical trials [350–352]. This will be described in more detail later and in a separate chapter by this group. Hsia originated a dialdehyde prepared from open ring sugars to form polyhemoglobin [200–201]. This has been developed and extended by Hemosol and is now being tested in phase II clinical trials [124]. This will be described later and in a separate chapter by this group. Many other bifunctional 2,3–DPG pocket modifiers are also being studied [275, 276]. As will be described later, a PEG conjugated bovine hemoglobin is now in phase I clinical trials [397]. Another very exciting area is the use of recombinant technology to produce recombinant human hemoglobin from *E. coli* [264, 265]. As will be discussed later, this has been developed by Somatogen and is being tested in phase II clinical trials [45, 295, 296]. This group will also describe this in a chapter in the second volume of this book.

Thus, crosslinked and recombinant hemoglobin is now at a very exciting stage with several ongoing clinical trials. In the future, there will be other new chemical modifiers including those proposed by Bucci et al [34–37], Kluger [250] and many others. New recombinant hemoglobins are also being developed, for instance by Fattor and Mathews [148], Fronteicelli [159], Olson [332–334] and many others. More details of specific examples will be given later.

```
┌─────────────────────────────────────────────────────────────────┐
│                    CROSSLINKED HEMOGLOBIN                          │
│                       SOME EXAMPLES                                │
│                                                                   │
│   CROSSLINKERS - INTERMOLECULAR & INTRAMOLECULAR                  │
│   Glutaraldehyde                                                  │
│   Dialdehydes from oxidized ring structure of sugars             │
│   Other polyaldehydes, diimidate esters etc                      │
│                                                                   │
│                                                                   │
│   CROSSLINKERS - INTRAMOLECULAR                                   │
│   2-Nor-2-formylpyridoxal 5'-phosphate                           │
│   Bis-(3,5-dibromosalicyl)furmarate                              │
│   Many others being explored                                     │
│                                                                   │
│                                                                   │
│   CARRIERS FOR CONJUGATED HEMOGLOBIN:                            │
│   Dextran-aldehyde                                               │
│   Polyethylene glycol                                            │
│   Polyxyethylene and others                                      │
│                                                                   │
└─────────────────────────────────────────────────────────────────┘
```

Fig. 16. Excellent in–depth studies to design specific crosslinkers and modifiers as shown here are being carried out by many groups. This has resulted in exciting improvements in the specificity of crosslinkers. Some of these also have the dual function of modifying the 2,3–DPG pocket of the hemoglobin molecule to improve the oxygen release properties.

Laboratory Procedure for the Preparation of Pyridoxalated Polyhemoglobin

For those entering this area of research for the first time, a major complaint is the lack of any books containing laboratory procedures for the preparation and testing of blood substitutes, procedures that they can readily use in their laboratory for preparing and testing modified hemoglobin. In view of this, the following is a simplified laboratory method for preparing pyridoxalated polyhemoglobin. This is the one we use in our laboratory for researchers starting in research on modified hemoglobin. This contains most of the basic laboratory principles and can form the basis of extension into new approaches. This is also good for preparing basic modified hemoglobin for use in different laboratory research. However, this is not for use in preclinical or clinical application. For this, more detailed procedures are required, especially in the preparation of ultrapure stroma–

free hemoglobin. Other important areas include stringent care in sterility; testing for endotoxin and complement activation; freedom from contaminating stroma fractions; and viral and bacterial inactivation. These are complicated and expensive procedures practical for use in larger industrial facilities and are not included here.

Preparation of Stroma–Free Hemoglobin Solution

The most important starting material, hemoglobin, has to be very carefully prepared. The easiest way is to buy this when it is available, for example, from Biopure in the form of purified bovine hemoglobin. Otherwise, it can be prepared in the laboratory as follows.

Out–dated whole blood obtained from blood bank or fresh animal blood is centrifuged at 6000 rpm (4000 g) for 20 minutes at 4°C. The plasma and buffy coat containing the white blood cells and platelets are removed by aspiration. The sedimented red blood cells are washed four times as follows. In each washing, three times their volume of ice–cold, sterile isotonic saline is added and mixed, then centrifuged to remove the saline. Red blood cells are then lysed by the addition of two volumes of hypotonic, 15 milliosmolar phosphate buffer, pH 7.4 to one volume of packed cells. This is mixed by repeated inversion and swirling for 2–3 minutes. After standing for 20–30 minutes, the suspension is placed into a large separation funnel and 0.5 volume of cold, reagent–grade toluene is added. This is mixed by shaking vigorously for five minutes, then left standing for three hours at 4°C. The top layer consisting of toluene with extracted stroma lipid and cellular debris is aspirated. The solution is then centrifuged at 25,000 g at 4°C for one hour. This is followed by a second toluene extraction but instead of left standing for three hours, it is left overnight in a cold room or refrigerator at 4°C. The lower layer of hemoglobin solution is then separated and centrifuged at 25,000 g at 4°C for one hour.

Although not absolutely required, a crystallization step can be carried out to improve the purity. Here, the hemoglobin is crystallized in a 2.8 M phosphate buffer as follows. The stroma–free hemoglobin is dialyzed (1:5 v/v) against 2.8 M potassium phosphate (pH 7.0) at 37°C for five hours, then for 18 hours against a fresh 2.8 M potassium phosphate buffer. The crystals from inside the dialysis fiber are re–dissolved for the next step.

The final step is to dialyze the stroma–free hemoglobin solution against Ringer's lactate solution 4°C.(An isotonic, buffered dialysate solution pH 7.35 can also be prepared using NaCl 5.85 g/L; NaHCO$_3$ 3.61/L; NaH$_2$ PO$_4$.H$_2$O 0.44 g/L; KCl 0.37 g/L; CaCl$_2$.2H$_2$O 0.22 g/L; MgCl$_2$.5H$_2$O 0.08 g/L). For small volumes, dialysis can be easily carried out by placing the solution in dialysis tubing. *Both sides of the tubing are carefully tied. The dialysis tubing is then suspended in the Ringer's lactate solution kept stirred with a magnetic stirrer. For larger volumes, a standard hollow fiber dialyzer is required.*

Pyridoxylation of Hemoglobin

This is based on a modification of the method of Benesch [20]. Stroma–free hemoglobin is deoxygenated under continuous nitrogen bubbling for 1–2 hours at 4°C. Pyridoxal–5'–phosphate (Sigma Chemical Co.) in Tris–HCl is added in a 4:1 molar ratio, then reduced with excess $NaBH_4$ under nitrogen for 18 hours. Excess reagents are removed by dialysis against Ringer's lactate. An isotonic, buffered dialysate solution pH 7.35 can also be prepared in the laboratory: NaCl 5.85 g/L; $NaHCO_3$ 3.61/L; $NaH_2\ PO_4.H_2O$ 0.44 g/L; KCl 0.37 g/L; $CaCl_2.2H_2O$ 0.22 g/L; $MgCl_2.5H_2O$ 0.08 g/L.

Preparation of Pyridoxalated Polyhemoglobin

This is carried out in a cold room at 4°C. First, 0.4 ml of 1.3 M lysine monohydrochloride in a 0.1 M phosphate buffer is added to 30 ml of pyridoxalated hemoglobin (10 g/dl). Next, 3 ml of ice–cold degassed 0.25 M glu- taraldehyde in 0.1 phosphate buffer is added slowly. The reacting mixture is left rotating in a mixer in the cold room. Methemoglobin formation will be marked if reaction is not carried out at 4°C. The progress of the reaction can be moni- tored either by agarose gel chromatography columns (Sepharose 4B – 60,000–2 million daltons and Sepharose 6B 10,000–2 million daltons) or by a colloid os- mometer. This gives information on the molecular distribution of the polyhemoglobin. A rough estimate of the reaction can be done by ultrafiltering a small sample in a small 100 dalton cutoff small centrifuge tube type of ultrafiltrator. The amount of tetramers ultrafiltered gives a quick estimate of the degree of polymerization of the hemoglobin preparation. The molecular weight range of the polyhemoglobin can be controlled by the time of reaction, ratio of glutaraldehyde to hemoglobin and other reaction parameters.

When the desired molecular weight range of polyhemoglobin is obtained, the crosslinking is quenched. This is done by adding 40 ml of the 1.3 M lysine monohydrochloride in a 0.1 M phosphate buffer. The solution is then centrifuged at 25,000 for one hour. The supernatant is dialyzed at 4°C for three hours against a Ringer's lactate solution or an isotonic, buffered dialysate solution pH 7.35 as described above. Dialysis removes free glutaraldehyde and excess lysine. It also allows the electrolytes to be adjusted to physiological concentrations. Removal of unreacted reagents is very important. To allow for better surface–volume relationship for adequate dialysis, only 10 ml is placed in each very partially filled tubing. Dialysis is completed when the electrolyte composition of the final product is the same as Ringer's lactate.

Although not needed for laboratory preparation, for better final purification all hemoglobin preparations are passed through a composite artificial kidney consisting of a hollow fiber dialysis in series with a 100 g column of collodion-coated activated charcoal. The collodion–coated activated charcoal can be prepared using the laboratory procedure as described [60].

The solution is ultrafiltered to the desired hemoglobin concentration. The solution is then filtered through a 0.45 micron sterilization filter unit. It is stored under sterile condition at 4°C to slow methemoglobin formation.

Sources of Hemoglobin

Where do we obtain all the hemoglobin needed for preparing modified hemoglobin? Besides hemoglobin from human sources, bovine hemoglobin is another possibility. Feola's original group has been the early pioneers in developing bovine hemoglobin [149]. Unlike human hemoglobin, bovine hemoglobin outside red blood cells has a high P_{50} without requiring 2,3–DPG or its analogs. Feola's group has carried out much of the earlier research on bovine hemoglobin for blood substitutes [149–151]. Bovine hemoglobin has been crosslinked into polyhemoglobin and is being used in clinical trials by Biopure [206–210, 222]. Bovine hemoglobin has also been used by Enzone as PEG-conjugated bovine hemoglobin [314, 315, 397, 398]. As discussed earlier, human hemoglobin produced by recombinant technology in microorganisms has already been developed and is now in phase II clinical trials [45, 148, 159, 264–265, 295, 296, 329, 332–334, 394]. The recombinant approach also allows the modification of hemoglobin to prevent dimer formation and to have a good P_{50}. Another potential approach that has not been fully developed is transgenic human hemoglobin in animals [254, 255, 327]. And another potential approach is Tsuchida's synthetic heme [428].

Blood Substitutes: Principles, Methods, Products and Clinical Trials, by Thomas Ming Swi Chang. © 1997 Karger Landes Systems.

•••••••••••••••••••••••••••

What Are the Functional and Efficacy Properties of Crosslinked Hemoglobin?

Composition

Being sure that the final modified hemoglobin preparation has the correct final composition is extremely important. The electrolyte composition has to be analyzed. Hemoglobin and methemoglobin concentrations are also important. Furthermore, for animal studies and clinical trials, sterility, endotoxin and other detailed characterizations are required. Modified hemoglobin preparations are prepared to have electrolyte composition similar to that of plasma. Having oncotic pressure (colloid osmotic pressure) close to that of plasma is also important. Most modified hemoglobin solutions have similar electrolyte composition and oncotic pressures. There are variations in the types of other chemicals added to the preparation for slowing methemoglobin formation, for antioxidant effects and for other actions.

The specific example below [236] is that of pyridoxalated polyhemoglobin prepared from the laboratory procedure described above. It must be remembered that this is prepared from a laboratory procedure without the more complicated methods of optimization.

Molecular Weight Distribution

Molecular weight distribution varies widely among the different types of crosslinked hemoglobin blood substitutes. For the intramolecularly crosslinked hemoglobin, they are all in the tetrameric form with a molecular weight of about 68,000. In polyhemoglobin the molecular weight distribution can vary very

Table 1. Composition of pyridoxalated polyhemoglobin
solution

Assays	Mean ± S.D.
Hemoglobin (g/dl)	*13.5 ± 1.0*
MetHb (g/dl)	*0.4 to 1.0 ± 0.12*
Na⁺ (mEq/L)	*145.0 ± 5.2*
K⁺ (mEq/L)	*4.9 ± 0.3*
Cl⁻ (mEq/L)	*113.6 ± 2.6*
Ca⁺⁺ (mg/dl)	*11.2 ± 1.0*
Mg⁺⁺ (mg/dl)	*2.0 ± 0.2*
Phosphate (mg/dl)	*3.2 ± 0.1*
Osmolality (mOsm/kg)	*312.4 ± 9.9*
pH (@ 37°C)	*7.37 ± 0.05*

widely. In polyhemoglobin, the molecular weights follow a distribution curve from tetrameric to much larger molecular weight. Some groups prefer using a larger mean molecular weight. Other groups prefer using a smaller mean molecular weight. Those with lower mean molecular weight also have more tetrameric hemoglobin. Those with higher mean molecular weight have less tetrameric hemoglobin because of the higher degree of polymerization. Gould's group has emphasized the need to remove as much tetrameric hemoglobin as possible [176]. After their preparation of polyhemoglobin, they carried out another step to remove much of the tetrameric hemoglobin. With such wide variation in molecular weight distribution from different groups, generalization is not possible. However, the pyridoxalated polyhemoglobin prepared from the laboratory procedure described can be used as an example. Even with this procedure, the molecular weight distribution can be varied at will by changing the reaction time, ratio of glutaraldehyde to hemoglobin and other reaction parameters.

For MW distribution, pyridoxalated polyhemoglobin (PP–PolyHb) was run on Sephadex G–200 (1.6 x 70 cm column) in 0.1 M Tris–HC1 (pH 7.5) at 12 ml/hr (Fig. 17). A selectivity curve (K_{av} vs. log MW) generated from elution volumes of three protein standards (thyroglobulin, catalase, aldolase, Pharmacia Fine Chemicals), was used to approximate molecular weights of the different fractions comprising the PP–PolyHb preparation. The K_{av} values for each peak enabled an approximation of the MW of each peak. Percent contribution of these fractions to overall composition of the PP–PolyHb solution in Table 2, was estimated from areas under each peak. The fraction of polymerized hemoglobin was 93%, with 60% falling in the 130,000–350,000 range and 33% ranging in molecular weight from 350,000–750,000. Only 7% was in the tetrameric form.

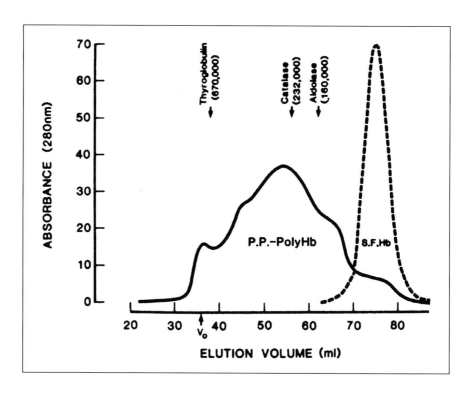

Fig. 17. Elution profile of pyridoxalated polyhemoglobin (P.P.–PolyHb) and stroma free hemoglobin (S.F.Hb) on Sephadex G–200(1.6 x 70 cm) column in 0.1 M Tris–HCl (pH 7.5) at 12 ml/hour [239]. Proteins of known molecular weights were used as markers. Reprinted with permission from Keipert PE, Chang TMS. Biomaterials, Artificial Cells and Artificial Organs 1988; 16:185–196. Courtesy of Marcel Dekker Inc.

Table 2. Molecular weight distribution of pyridoxalated-polyhemoglobin

Peak	V_e (ml)	K_{av} (daltons)	M.W.	Percent
1st	37.7	0.02	750,000	10
2nd	45.5	0.11	470,000	23
3rd	54.5	0.22	260,000	39
4th	65.4	0.35	130,000	21
5th	77.2	0.49	66,000	7

NOTE: $K_{av}=(V_e-V_o)/(V_t-V_o)$ where V_e-elution volume, V_o-void volume, and V_t=bed volume.

The average molecular weight of the preparation can be decreased by simply decreasing the time of reaction or by decreasing the molar ratio of glutaraldehyde:hemoglobin. For example, decreasing this ratio by 50% will markedly decrease the mean molecular weight with a very larger fraction of tetrameric hemoglobin (both crosslinked and free).

Oncotic Pressure

Oncotic Pressure Across the Capillaries

Fluid movement across the capillaries is the result of two forces:

1. Hydrostatic pressure is due to the blood pressure from the pumping action of the heart and the vascular resistance. The net effect of hydrostatic pressure itself across the capillary is the movement of fluid out of the capillary.

2. Oncotic pressure. Osmotic pressure across a membrane is exerted by an impermeable or unequilibrated solute. Under physiological conditions, plasma proteins are the only plasma solutes that cannot move quickly across the capillary wall. Plasma proteins therefore exert a colloid osmotic pressure (oncotic pressure) across the capillary wall. Under physiological conditions, the net effect of oncotic pressure itself is the movement of fluid into the capillary.

The protein concentration of plasma is about 7 g/dl. Hemoglobin inside red blood cells or microencapsulated hemoglobin is separated from the plasma and not dissolved in the plasma. They therefore do not exert an oncotic pressure. However, free hemoglobin when infused as a solution, will behave like protein dissolved in the plasma. Thus a 7 g/dl free hemoglobin solution will have about the same oncotic pressure as plasma – isooncotic with plasma.

The concentration of numbers of soluble particles decides the amount of oncotic pressure. Each free hemoglobin molecule behaves like one solute particle. A polyhemoglobin (PolyHb) containing four hemoglobin molecules would behave as one solute particle. This one PolyHb solute particle therefore exerts less oncotic pressure than four free hemoglobin molecules. However, a polyhemoglobin solution is made up of solute particles ranging from one to four or more hemoglobins in each particle. Thus depending on the composition, the oncotic pressure can vary. Furthermore, there is the effect from macromolecular–water interaction of large molecules. Generally speaking, a 14 g/dl polyhemoglobin solutions exerts the same oncotic pressure as 7 g/dl of free hemoglobin. They are therefore isooncotic with plasma. Modified hemoglobin solutions are usually prepared to be isooncotic with plasma. A specific example is as follows (Fig. 18):

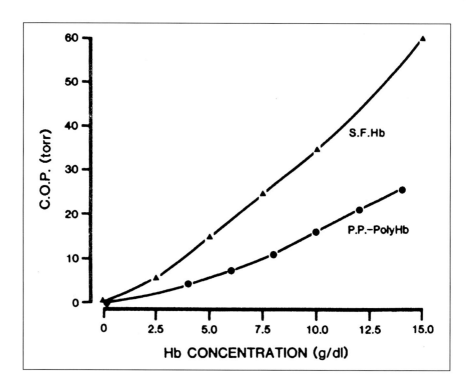

Fig. 18. The relationship between hemoglobin (Hb) concentration and oncotic pressure (colloid osmotic pressure C.O.P) expressed as torr (mmHg) [239]. Several Hb molecules are linked to form one polyHb solute particle. Oncotic pressure depends roughly on the numbers of solute particles. As a result, for the same amount of Hb, polyHb has less [sic] solute particles than stroma–free hemoglobin (Hb) and thus exerts less oncotic pressure. Reprinted with permission from Keipert PE, Chang TMS. Biomaterials, Artificial Cells and Artificial Organs 1988; 16:185–196. Courtesy of Marcel Dekker Inc.

Oncotic Pressure of a Laboratory Prepared Polyhemoglobin Solution

We will continue to use the pyridoxalated polyhemoglobin prepared from the laboratory procedure described here as a specific example. Oncotic pressure (Colloid osmotic pressure COP) can be obtained with an oncometer (e.g., IL 186 Weil Oncometer calibrated with 5 g% albumin solution). The COP of both pyridoxalated polyhemoglobin (PP–PolyHb) and stroma–free hemoglobin (SFHb) was measured at various Hb levels. The effect of Hb concentration and the effect of crosslinking on colloid osmotic pressure (COP) can be seen in Fig. 18. To attain isooncotic pressures (COP=25 mmHg) SFHb had to be diluted to 7 g/dl, compared with crosslinked PP–PolyHb at 14 g/dl. The relationship between COP and Hb concentration was slightly curvilinear. The COP values for SFHb were at least double that of PP–PolyHb at similar concentrations (Fig. 18).

Viscosity

Viscosity varies with the type of modified hemoglobin, the concentration of the modified hemoglobin, and for polyhemoglobin, the molecular size of the polyhemoglobin. The composition of the suspending medium is sometime also important. Again, using the laboratory example of pyridoxalated polyhemoglobin, these were measured in an Ubbelohde Viscometer (Canon Instrument Co.) equilibrated at 25°C and at 37°C. Multiplying kinematic viscosity times, the measured density of each solution yielded values for dynamic viscosity (centiPoises). For the 14 g/dl pyridoxalated polyhemoglobin prepared with the above molecular weight distribution, the intrinsic (dynamic) viscosity is 3.83 cP at 37°C that is similar to that of whole blood. Lower molecular weight distribution would result in lower viscosity.

Oxygen Affinity and Cooperativity

General

Oxygen affinity is a measure of the ease of unloading of oxygen to the tissue as it passes through the capillaries. Hemoglobin binds oxygen from the lung. As it reaches the tissues, hemoglobin releases oxygen to the tissue depending on the oxygen tension in the tissue. Oxygen affinity is a measurement of how easily oxygen can be released from hemoglobin when required by the tissues. Without 2,3–DPG, crosslinked hemoglobin cannot readily release oxygen when required. Pyridoxal phosphate (PLP) applied to hemoglobin can replace 2,3–DPG [20]. However, Greenburg et al showed that the PLP–hemoglobin breaks down into dimers in the circulation [177]. Therefore pyridoxalation was combined with polyhemoglobin and conjugated hemoglobin. Recent studies include the use of bifunctional crosslinkers that also modified the 2,3–DPG pocket. This has further improved oxygen release characteristics. At present, modified hemoglobin can be prepared with better oxygen releasing properties than hemoglobin inside red blood cells. However, one cannot increase the oxygen releasing properties indefinitely. There is a point beyond which the ability of hemoglobin to bind oxygen from the lung will be compromised. P_{50} varies markedly among the different types of modified hemoglobin. The new type of modified hemoglobin with specific modification of the 2,3–DPG pocket has very high P_{50} of more than 30 mmHg.

A Specific Example

The following is a specific example, again using pyridoxalated polyhemoglobin prepared from the laboratory procedure described [237]. This is used mainly as an example for simple reproduction in the laboratory; thus P_{50} has not been optimized. Oxygen dissociation curves are measured under standard conditions (pH = 7.4, pCO_2 = 40 mmHg, temp = 37°C) using tonometry, an IL 282 Hb co-oximeter, and a Corning 175 pH/Blood Gas Analyzer. As discussed earlier, P_{50} measures the ease of hemoglobin to unload oxygen at a given oxygen tension. P_{50} is the oxygen tension at which fully oxygen saturated hemoglobin releases 50% of the oxygen it carries. In addition, as discussed earlier, cooperativity of hemoglobin allows it to carry oxygen more efficiently. Cooperativity is measured using the Hill coefficient (n–value).

Using these laboratory measurements in this specific example, whole blood in this case has a P_{50} of 24–26 mmHg and a Hill coefficient of n = 2.51. Stroma–free hemoglobin has a lower P_{50} of 12–14 mmHg because of the absence of 2,3–DPG. It has a Hill coefficient of n = 2.48. For pyridoxalated hemoglobin before crosslinking into polyhemoglobin, the P_{50} is 28 torr and, Hill coefficient of n = 2.5. Here, the pyridoxal 5'–phosphate, acting as a 2,3–DPG analog, effectively increased the P_{50} of stroma–free hemoglobin. After polymerization, pyridoxalated polyhemoglobin has a P_{50} of only 20 mmHg and a Hill coefficient of n = 1.88. (With special optimization P_{50} can be increased to 28 mmHg). Glutaraldehyde crosslinking, as described in this laboratory example, has decreased the P_{50} of pyridoxalated hemoglobin and restrict subunit interactions necessary for cooperative oxygen binding. This is shown by the loss of the sigmoidal character of the oxygen dissociation curve at the lower pO_2 values that corresponded to a lower n–value (Hill coefficient) of 1.88. The net effect of pyridoxalation and polymerization of stroma–free hemoglobin is to improve its P_{50} from 12–26 mmHg to 20 mmHg, or higher with optimization. This is at the expense of a decrease of cooperativity as measured by the Hill coefficient of n = 2.48 to 1.88. More complicated special reaction conditions can improve the P_{50} to 28 mmHg.

Unmodified Hemoglobin

As discussed earlier, after infusion hemoglobin is broken down from its tetrameric form into two dimers. These dimers are filtered out into the kidney and cause renal toxicity. If all hemoglobin molecules in the infusion are modified hemoglobin (intermolecular, intramolecular, conjugated or recombinant) that do not enter the kidney, then they should not be toxic to the kidney. Unfortunately, it is not possible to remove all the unmodified free hemoglobin molecules. Thus safety requirements depend on setting a limit on the concentration of unmodified free hemoglobin in the modified hemoglobin preparations.

Methemoglobin Formation

General

The present first generation modified hemoglobin is prepared from ultrapure hemoglobin. As such, it does not contain reducing enzyme systems normally present in red blood cells. Therefore in solution, modified hemoglobin tends spontaneously to oxidize to methemoglobin that cannot carry O_2. Different approaches are used to slow methemoglobin formation. Methemoglobin formation is slower when hemoglobin is in the deoxyhemoglobin form. Methemoglobin formation is also slower when hemoglobin carries carbon monoxide. Other methods include the use of higher pH, the addition of specific reducing agents or antioxidants. The following is a specific example using lyophilization for the polyhemoglobin prepared from the laboratory procedure [239].

Lyophilization

Lyophilization of pyridoxalated polyhemoglobin (PP–PolyHb) solution results in a 10–fold increase in MetHb. One way to prevent this is to add 3.0% glucose before freeze–drying. The protective effect of the glucose is shown by post–lyophilization MetHb levels that are only 4.6 ± 1.9% higher than the initial levels [239]. Storage as a freeze–dried powder at 25 °C is feasible for six months after which MetHb formation increased to 9.4% per year (Fig. 19). However, if lyophilized PP–PolyHb is kept at –20 °C long–term storage is possible. Here MetHb levels remains below 6% after three years. This corresponds to an average rate of MetHb formation of 0.5% per year (Fig. 19).

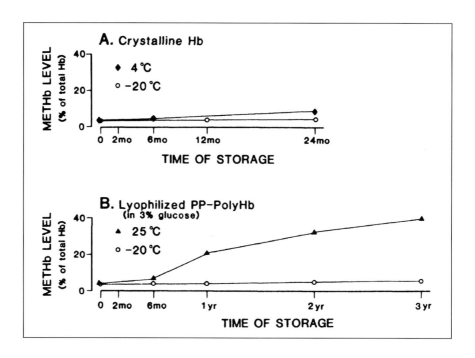

Fig. 19. Methemogloin (METHb) formation [239]. Crystalline hemoglobin (Hb). Lyophilized pyridoxalated polyhemoglobin (PP–PolyHb). Reprinted with permission from Keipert PE, Chang TMS. Biomaterials, Artificial Cells and Artificial Organs 1988; 16:185–196. Courtesy of Marcel Dekker Inc.

Experimental Hemorrhagic Shock Model

Crosslinked hemoglobin is effective in the resuscitation of hemorrhagic shock [84, 191, 237, 272, 305, 321]. This is as effective as whole blood with 100% long–term recovery in unanesthetized animals that had lost 67% of their total blood volume. Conjugated hemoglobin is also effective in hemorrhagic shock [2, 220, 280, 314, 387]. Microencapsulated hemoglobin is also effective in hemorrhagic shock [372, 435].

Hemorrhagic Shock Model

When a small volume of blood is lost, the use of fluid to replace the volume is enough to maintain blood pressure and perfusion. It is only when a very large volume of blood is lost that red blood cell replacement is needed (Fig. 20).

```
┌─────────────────────────────────────────────────┐
│ HEMORRHAGIC SHOCK MODELS                          │
│ (1)  REMOVAL OF 30-40% BLOOD VOLUME               │
│        • needs volume expander                    │
│        • does not need rbc replacement            │
│                                                   │
│ (2)  REMOVAL OF 67% BLOOD VOLUME                  │
│        • needs volume expander                    │
│        • needs rbc replacement                    │
└─────────────────────────────────────────────────┘
```

Fig. 20. Hemorrhagic shock models to assess red blood cell (rbc) replacement. Removal of 30–40% blood volume requires mainly fluid replacement and therefore not the best model for rbc replacement. The removal of 67% blood volume is a better animal model for rbc replacement. Reprinted with permission from Chang TMS, Biomaterials, Artificial Cells and Immobilization Biotechnology 1992; 20:154–174. Courtesy of Marcel Dekker Inc.

Messmer has discussed in some detail volume replacement in hemorrhagic shock [290]. Solutions for volume replacement include saline, Ringer's lactate, Ringer's lactate–Dextran, hypertonic saline and others. Most of these studies are based on bleeding 30–40% of the blood volume. These models are excellent for the study of volume replacement when red blood cell replacement is not conclusively needed [452]. This model is therefore not sensitive enough for testing the efficacy of red blood cell substitutes in hemorrhagic shock [97].

Therefore, to test the effectiveness of blood substitutes for volume and red blood cell replacement, a more severe model is needed. A modification of the 1950 model of Wiggers is suitable for this. This involves two stages of bleeding resulting in the removal of 2/3 of the total blood volume. We modified this using the chronic tail–cannulation technique [81] to form an unanesthetized model [237]. Another group used another chronic cannulation technique [272] that works as well. Using these unanesthetized models, crosslinked hemoglobin blood substitutes are shown to be more effective than standard volume replacement [237], 3 volume Ringer's lactate [84, 272] or 7.5% hypertonic–saline/6% dextran 70 plus 3 volumes Ringer's lactate [84].

```
┌─────────────────────────────────────────────────┐
│ EXPERIMENTAL DESIGNS                             │
│ (1) Replaced by single infusion                 │
│ (2) Repeated infusions to maintain BP           │
│ (3) Acute studies – few hours                   │
│ (4) Chronic studies – follow recovery           │
│ (5) Others                                       │
└─────────────────────────────────────────────────┘
```

Fig. 21. Different experimental designs for hemorrhagic shock studies. Reprinted with permission from Chang TMS, Biomaterials, Artificial Cells and Immobilization Biotechnology 1992; 20:154–174. Courtesy of Marcel Dekker Inc.

Experimental Designs

Different experimental designs have been used (Fig. 21). Some experimental designs are for a single blood replacement. They are then followed without other treatment or monitoring until recovery or death [84, 237]. This is more comparable to emergency replacement in major disasters or wars in the field outside the hospital setting with no constant monitoring. Other experimental designs are for repeated infusion with continuous monitoring to maintain blood pressure. This is more applicable to surgical or hospital settings. Some designs are more suitable for battlefield injuries [191, 272]. Furthermore, some experiments are acute experiments of a few hours. Here, only the vital signs are continuously monitored. Obviously, the demonstrated effectiveness of blood substitutes will vary depending on the experimental design.

Not making general conclusions regarding the effectiveness of blood substitutes in hemorrhagic shock or in other applications is important. A given blood substitute is only as effective as the experimental design for which it has been tested. Some applications only require short–term effectiveness with constant monitoring. Other applications may require longer–term effectiveness without the availability of constant monitoring or medical care.

Typical Example of Experimental Study on Hemorrhagic Shock

Animal Model

Sprague–Dawley rats, 340 ± 40 gm, from Charles River were randomly divided in groups of six (n = 6). They were allowed to acclimatize for four days. They then received chronic cannulation as referred to above. During the 6–8 days of the postsurgical period, the catheters were flushed daily with sterile heparinized saline. Body weights and hematocrits were followed and only those with the following values 6–8 days after surgery were used. This included average body weight of 361.14 ± 10.53 gm. and hematocrit of 37 ± 2.7%.

Fully conscious rats were partially restrained allowing free access to the tails and catheters. Each rat received intravenous heparin, 150 IU/kg. One arterial cannula was connected to a statham pressure transducer for blood pressure and heart rates. These were recorded continuously on a Grass polygraph recorder. After steady baseline recording, lethal hemorrhagic shock was induced as follows. This involved the removal of 67% of the total blood volume through the other arterial catheter at 0.5 ml/min. This was in two stages. Removal of 36% of total blood volume was followed by a 10 min rest period. The second stage involved the removal of 31% of total blood volume. Throughout the bleeding, the blood pressure was maintained between 40 and 60 mmHg. 67% of total blood volume was removed from each rat with an average time of 46.52 ± 2.71 min.

Resuscitation Solutions

There were eight groups of rats with six rats randomly assigned to each group.
1. *Controls – no resuscitation fluid.*
2. *Reinfusion of rats' own shed whole blood.*
3. *Ringer's lactate solution. The volume was equivalent to the volume of shed blood.*
4. *Ringer's lactate solution equivalent to three times the volume of shed blood.*
5. *Sterile human albumin 7 gm% in Ringer's lactate The volume was equivalent to the volume of shed blood.*
6. *Stroma–free hemoglobin in Ringer's lactate solution. The volume was equivalent to the volume of shed blood.*
7. *Polyhemoglobin or other blood substitutes in Ringer's lactate solution. The volume was equivalent to the volume of shed blood.*

8. 4 ml/kg (7.5 g% NaCl/7 g% Dextran 70 followed by three times the volume of shed blood of Ringer's lactate.

Immediately after shock induction, each rat received one of the above intravenous resuscitation fluids at a rate of 0.5 ml/min. Mean arterial pressure and heart rates were recorded up to 30 minutes after the completion of infusion. Each rat was then returned to a separate cage. They were monitored for 14 days. Those rats that lived up to 14 days were considered as having "survived." The time of death was recorded for those that died before this time.

Acute Response

The short-term response of the animals is shown in Fig. 22. In the control group that did not receive any replacement, the blood pressure fell so that 30 minutes after the completion of the 67% bleeding, all the animals have died (Fig. 22). For those three groups that have received one volume or three volumes of Ringer's lactate or 7 g% albumin in Ringer's lactate, the blood pressure did not return to the control level (Fig. 22). The animals were maintained at this lower level for about 60 minutes. Hypertonic saline/dextran followed by three volumes Ringer's lactate sustained the blood pressure only in the first 10 minutes. Whole blood, stroma–free hemoglobin, and polyhemoglobin maintained the blood pressure at control levels (Fig. 22) throughout the one-hour period.

Long-Term Effects

The results of long-term survival (Fig. 23) show that only whole blood and polyhemoglobin are effective for the long–term recovery of the animals. One objective of this approach is to study the effect of a single transfusion on the long–term survival of the animals. This is one transfusion followed by no other special medical care or other types of treatment. All the animals followed the standard animal center diet and fluid by mouth. If we did not follow long-term effects, the short–term response (Fig. 22) would have shown that stroma–free hemoglobin is as effective as polyhemoglobin.

General

The above animal model of lethal hemorrhagic shock conclusively shows the effectiveness of modified hemoglobin blood substitutes. Rat models are useful to study survival rates using the necessary controls. The major disadvantage

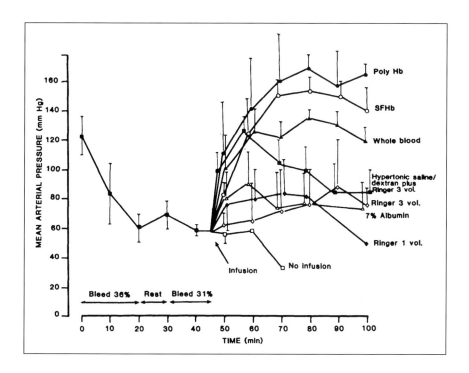

Fig. 22. Lethal hemorrhagic shock rat model. Results of a single replacement using one of polyhemoglobin (PolyHb), stroma–free hemoglobin (SFHb) or other solutions. Reprinted with permission from Chang TMS, Varma R. Biomaterials, Artificial Cells and Immobilization Biotechnology 1992; 20:503–510. Courtesy of Marcel Dekker Inc.

is that in such small animals taking blood samples for blood biochemistry and hematology is difficult. This is especially true in the present study where the major aim is to follow the long–term survival of a lethal hemorrhagic shock model.

As will be discussed later, using this type of lethal hemorrhagic shock model in clinical trial in human is not possible, since lethal controls in clinical studies is out of the question. Thus, the effectiveness for hemorrhagic shock will be much more difficult to show in clinical trials. Several alternatives have been suggested. For example oxygen transport parameters, central venous oxygen tension, acid–base balance, lactic acid and other parameters [15, 457, 469]. However, based on our earlier discussions, we have to be sure that the patients studied this way have lost enough blood to need red blood cell replacement.

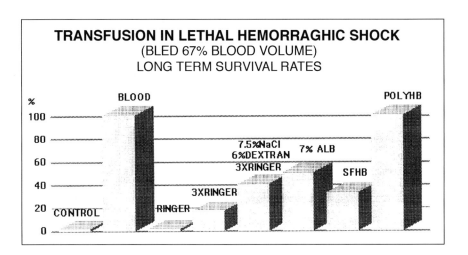

Fig. 23. Lethal hemorrhagic shock rat model. The same study as in the above figure, except that survivals were followed for 14 days after a single replacement using one of the solutions. Only whole blood and polyhemoglobin resulted in long term survival. Reprinted with permission from Chang TMS, Varma R. Biomaterials, Artificial Cells and Immobilization Biotechnology, 20:503–510, 1992. Courtesy of Marcel Dekker Inc.

Other Experimental Efficacy Studies Including Exchange Transfusion

General

Crosslinked hemoglobin can replace up to more than 95% of the total blood volume in experimental animal studies. It is therefore used for normovolemic hemodilution in elective surgery and other forms of major surgery. Other applications include its use in ischemic diseases, cardioplegia and others. The use and effects of modified hemoglobin in ischemic organ perfusion and in organ preservation are also being intensively investigated by many groups [3, 5, 142, 199, 213, 280, 307, 349, 418]. This includes those of Agishi et al, Alayash et al, Faassen et al, Horiuchi et al, Ilan et al, Matsushita et al, Mouelle et al, Ogilby et al, Pristoupil et al, Tanaka et al and others.

Exchange Transfusion

This is one of the frequently used experimental procedures to test the efficacy of a given blood substitute in normovolumic hemodilution. Again, we will use the laboratory prepared polyhemoglobin as a specific example [238]. Anesthetized male Sprague–Dawley rats with an average weight of 310 g were used. Cannulation of the femoral artery and femoral vein using 3 cm length PE10 tubing (0.28 mm inner, 0.61 mm outer diameter, Clay Adams) connected into a 20–cm length of PE50 tubing (0.58 mm inner, 0.97 mm outer diameter). These are filled with heparin (500 USP units/ml). Both catheters were connected to a pump (for example Minipulse–2 Gelson Medical Electronics) fitted with tubing (0.89 mm inner). Rats were heparinized before exchange transfusion by giving intravenously 60 units/100 g body weight. The polyhemoglobin solution is warmed to body temperature of 37°C and infused intravenously into the femoral vein. Simultaneously blood from the animal is removed from the femoral artery. The flow rate is maintained at 0.5 ml/min by adjusting the peristaltic pump. This is monitored by measuring the volume infused and the volume collected. Vital signs, hemodynamics and other measurements can be carried out as required.

Control solutions and test solutions are used in different groups of animals. For instance, stroma–free hemoglobin solution, bovine serum albumin, Ringer's lactate, homologous whole blood, and blood substitute. Isovolemic exchanges are in the ranges of 30%, 50%, 70%, 90% and 97%. Vital signs remained stable during isovolemic exchanges of homologous blood and pyridoxalated polyhemoglobin during these ranges. Circulation half times depends on the volume of polyhemoglobin exchanged. Thus, for the 97% exchange the half time is 30 hours as compared with 2.5 hours for stroma–free hemoglobin. For 15% exchange, the circulation half–time of polyhemoglobin is 7 hours compared with 1 hour for stroma–free hemoglobin.

End Point of Efficacy in Clinical Trial

As discussed earlier, animal studies have shown that modified hemoglobin is effective in lethal hemorrhagic shock of 67% blood loss and in exchange transfusion of up to 97% total blood volume. For obvious reasons, this type of severe test of efficacy using lethal animal models cannot be used in human clinical trials with lethal controls. As a result a different set of end points has to be used for clinical trials of efficacy in humans [155, 452, 468]. The FDA has published a summary on "points to consider" for efficacy designs in efficacy clinical trials in humans [155]. Various groups have used different approaches in their clinical studies. Their end–points include the ability of the product to maintain blood pressure, to maintain hemoglobin concentration or to decrease need of blood replacement. These will be discussed under the section on clinical trials. Winslow and his collaborators have worked extensively on oxygen transport and adequacy of oxygen perfusion as possible criteria – for example, Intaglietta et al [217, 218], Vandergriff et al [438, 439] and Winslow [455, 461].

Chapter 5

Blood Substitutes: Principles, Methods, Products and Clinical Trials,
by Thomas Ming Swi Chang. © 1997 Karger Landes Systems.

••••••••••••••••••••••••••••
How Safe Are Modified Hemoglobins?

General

In this regard, proper selection of the toxicity model is important. The readers are especially referred to FDA's "Points to consider in the safety evaluation of hemoglobin–based oxygen carriers" [154]. Animal studies using properly prepared polyhemoglobin have not shown adverse effects on coagulation factors, leucocytes, platelets or complement. Detailed immunological studies were also carried out. Homologous polyhemoglobins from the same species (e.g., rat polyhemoglobin injected into rats) are not immunogenic even with repeated injections. Heterologous polyhemoglobin (e.g., nonrat hemoglobin injected into rats) does not result in immunological response initially. The heterologous form only results in immunological response after repeated injections. Conjugation and microencapsulation markedly decreased the antigenicity of heterologous polyhemoglobin. Since hemoglobin has a high affinity for nitric oxide, extensive studies have been carried out to see the effects on vasoactivity. Another important area of safety study is the distribution of modified hemoglobin after infusion. Crosslinked hemoglobin does not cause suppression of the reticuloendo–thelial system. More detailed discussion follows.

Effects on Coagulation, Platelets, Leucocytes and Complement

Coagulation Factors and Platelet Activation

The effects on coagulation factors have been studied by many groups. Typical published results are as follows [317–321]. Albumin- and saline-infused rats were controls. The infusion volume was 10% of the rat's blood volume. The concentrations of polyhemoglobin in this study were 7 g/dl. Measurements for prothrombin time (PT) and activated partial thromboplastin time (PTT) were at 5 minutes, 2, 6, 24 and 72 hours after infusion. Factor X, fibrinogen, plasminogen, antithrombin III and antiplasmin were followed at 24 and 72 hours after infusion. Compared with saline control PT and PTT did not change significantly after hemoglobin infusion. There were no changes in Factor X, fibrinogen, antithrombin III and antiplasmin. There was no significant difference in plasminogen levels between the polyhemoglobin infused groups and the control saline group. The effect of pyridoxalated polyhemoglobin and stroma–free hemoglobin on ADP–induced platelet aggregation was also studied [257]. In this study, we used rat platelet–rich plasma (PRP) and light transmission aggregometry. We analyzed the effects of stroma–free hemoglobin (SFHb) and pyridoxalated polyhemoglobin (P–PolyHb) on platelet aggregation. P–PolyHb and SFHb solutions did not initiate or facilitate platelet activation.

Effects on Complement Activation, Leucocytes and Platelets

The effects of highly purified stroma–free hemoglobin and polyhemoglobin on C_3 and C_{3a} levels and on blood cell counts were studied [317, 318, 320]. Plasma samples from rats were incubated with hemoglobin solutions or control solutions. There were no significant differences in C_3 or C_{3a} levels between the plasma incubated with polyhemoglobin solutions and the saline control. However, endotoxin and membrane stromas significantly increased C3a and lowered C3. This study suggests that purified stroma–free hemoglobin and polyhemoglobin do not activate complements. On the other hand, contaminants like membrane stroma or endotoxin activate the complement system [150,151]. Based on these studies, using highly purified hemoglobin for preparing crosslinked hemoglobin is important.

Polyhemoglobin and stroma–free hemoglobin did not cause significant changes in total leucocyte, differential, or platelet counts [317, 318, 320, 322]. There was also no significant effect on ADP–induced platelet aggregation.

Biodistribution and Reticuloendothelial System

Early biodistribution studies of isotope-labeled-modified hemoglobin show that when polyhemoglobin and encapsulated hemoglobin are small enough to pass through the lung capillaries, their uptake is mainly by the reticuloendothelial system [49, 59]. Several groups have carried out detailed biodistribution studies of crosslinked hemoglobin. Some examples included those of Anderson et al [10]; Hsia et al [202]; Keipert et al [240]; Kim et al [243]; Marks et al [273] and others. Crosslinked hemoglobins do not result in reticuloendothelial suppression [273].

Immunology

General

Hemoglobin is only weakly antigenic. However, antibodies can be successfully produced to specific hemoglobin moieties, but the animals had to be boosted with large amounts of antigen to yield these results. Polymerization of protein is generally thought to enhance its immunogenicity. In order to test this in 1986 we started to analyze the in–vivo immunogenicity of crosslinking hemoglobin into polyhemoglobin [72, 75, 86, 190]. Estep's group has reported on diaspirin crosslinked hemoglobin [141]. More recently, Bakker's group also studied the immunology of polymerized hemoglobin [26].

Antigenicity Measured as Antibody Titers

The first basic study is to see whether crosslinking hemoglobin into polyhemoglobin affects antigenicity. Rats are immunized with either rat stroma–free hemoglobin or rat polyhemoglobin in Freund's adjuvant. To account for cross–species reactivity, two other groups of rats are immunized with either human stroma–free hemoglobin or human polyhemoglobin. Rat polyhemoglobin ranges in molecular weight from 600,000 to 4 million daltons. The heterologous polyhemoglobin for human polyhemoglobin has a molecular weight range of 130,000–600,000. Antigens were labeled with ^{125}I. Percent protein bound radioactivity ranged from 87–97%.

As can be seen from the antibody titers in Fig. 24 rat hemoglobin (homologous) is not antigenic when injected with Freund's adjuvant into rats. Even after rat hemoglobin is polymerized into large molecular weight polyhemoglobin, it

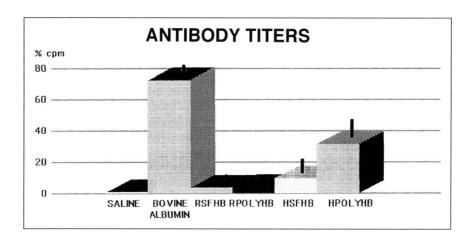

Fig. 24. Antibody titers after immunization in rat using Freund's adjuvant containing one of: Saline (negative control), Bovine albumin (positive control), RSFHb (rat stroma–free hemoglobin), RPOLYHB (rat polyhemoglobin), HSFHB (human or heterologous SFHb) and HPOLYHB (human or heterologous POLYHB). Reprinted with permission from Chang TMS, Varma R. Biomaterials, Artificial Cells and Artificial Organs 1988; 16:205–215. Courtesy of Marcel Dekker Inc.

is still not antigenic when injected into rat, even by this very severe test using Freund's adjuvant. On the other hand, hemoglobin from another species (heterologous) is antigenic and resulted in a significant increase in antibody titer. After crosslinking into polyhemoglobin, hemoglobin from another species is even more antigenic when injected into rats. However, the results also support the prior observation that hemoglobin is low in antigenicity, as can be seen when compared with heterologous albumin. Even heterologous polyhemoglobin showed lower antigenicity when compared with albumin. (Fig. 24). This low antigenicity agrees with previous results that showed hemoglobin to be weakly antigenic. Furthermore, the effect of crosslinking on increasing immunogenicity is only apparent under heterologous conditions. The results discussed are for the laboratory prepared pyridoxalated polyhemoglobin [72, 75, 190]. We have carried out similar studies using o–raffinose polyhemoglobin141. Similar results were obtained (Fig. 25).

Thus, the data suggest that within the same species, crosslinking of hemoglobin does not increase its antigenicity. On the other hand, repeated subcutaneous injections of heterologous hemoglobin or polyhemoglobin in Freund's adjuvant resulted in antibody production.

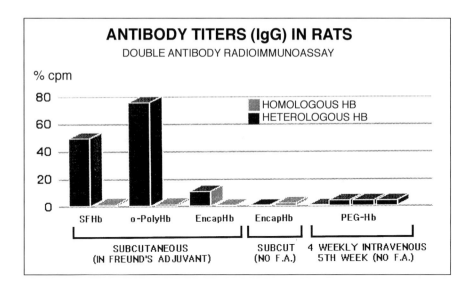

Fig. 25. Results of antibody titers (IgG) in rats with different types of hemoglobin (Hb). Stroma–free (SF) Hb, PolyHb; liposome encapsulated (Encap) Hb given subcutaneously with Freund's adjuvant (F.A.). Encap Hb give subcutaneously with no F.A.. Weekly PEG–conjugated bovine Hb given intravenously 5 times. Reprinted with permission from Chang TMS, Lister C, Nishiya T, Varma R. Biomaterials, Artificial Cells and Immobilization Biotechnology 1992; 20:611–618. Courtesy of Marcel Dekker Inc.

Effects of Infusion

Based on the above antibody titer results alone, we cannot make any conclusions regarding the immunological effect of the infusion of homologous or heterologous crosslinked hemoglobin. Further studies as described below need to be carried out [72, 75, 86].

First, four groups of rats that have not received any hemoglobin infusion before were used. Each group received one of the following infusions equivalent to 30% total blood volume: rat hemoglobin; rat polyhemoglobin; hemoglobin from another species (e.g., human); and polyhemoglobin from another species (human). Infusion of any one of the four solutions did not result in any adverse effects (Fig. 26) [72, 75].

The next group of study was carried out in immunized rats as follows [72, 75]. Four groups of rats each received subcutaneously immunizing doses of one of the above four solutions in Freund's adjuvant. After this, the antibody titers were analyzed. The rats in each group then receive infusions of the corresponding

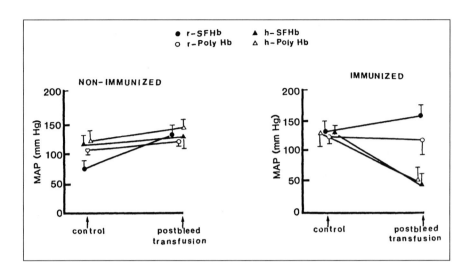

Fig. 26. Bleeding 30% blood volume in rats then replaced with the same volume of one the following: r–SFHb (rat stroma–free hemoglobin), r–Poly Hb (rat polyhemoglobin), h–SFHb (human or heterologous SFHb) and hPoly Hb (human or heterologous Poly Hb). Nonimmunized means rats have never been exposed to the solution. Immunized means rats had been immunized with the material in Freund's adjuvants. Reprinted with permission from Chang TMS, Varma R. Biomaterials, Artificial Cells and Artificial Organs 1988; 16:205–215. Courtesy of Marcel Dekker Inc.

hemoglobin solution. The volume is 30% of the total blood volume. Infusion of rat hemoglobin or rat polyhemoglobin did not result in any adverse effects (Fig. 26). However, infusion of hemoglobin or polyhemoglobin from another species into the immunized rats resulted in severe anaphylactic reactions (Fig. 26). The animals were in severe respiratory distress and gasping indicating severe bronchospasm. Furthermore, the blood pressure fell to shock levels (Fig. 26). The lung histology shows anaphylactic reactions in immunized rats that received the corresponding hemoglobin or polyhemoglobin from another species. The reaction is more severe in those immunized rats receiving heterogenous hemoglobin than polyhemoglobin. Eosinophils, mast cells and histiocytes accumulate in the bronchioles. Alveolar septa are thickened with histiocytes and mast cells. Perivascular edema with accumulation of histiocytes and mast cells are also observed. In addition, there are congestion, focal hemorrhage and focal atelectasis.

The above studies were repeated later using o–raffinose polyhemoglobin and the same results were obtained [86].

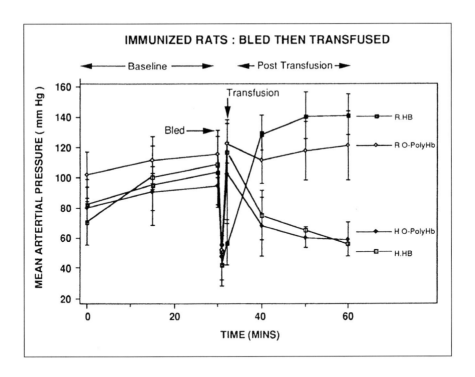

Fig. 27. Rats were immunized with Freund's adjuvant containing one of the following: RHb (rat hemoglobin), R.O–PolyHb (rat o–raffinose polyhemoglobin), H.Hb (human or heterologous Hb), H.O–PolyHb (human or heterologous o–raffinose polyhemoglobin). This figure shows the effects of replacing 30% bled blood volume bleeding with infusion of the corresponding hemoglobin solution. Reprinted with permission from Chang TMS, Lister C, Nishiya T , Varma R. Biomaterials, Artificial Cells and Immobilization Biotechnology 1992; 20:611–618. Courtesy of Marcel Dekker Inc.

The above infusion studies show that infusion of polyhemoglobin from the same species either for the first time or under the most severe immunization condition did not result in any adverse response (Fig. 28). Infusion of polyhemoglobin from a different species for the first time also did not result in any adverse response. However, when polyhemoglobin from a different species is infused after the most severe immunization procedures, there were severe anaphylactic reactions (Fig. 28). Thus, the result seems to show that polyhemoglobin from the same species would not have any immunological problems. It also shows that polyhemoglobin from a difference species also did not produce adverse effect when infused for the first time. What one cannot conclude from these results is how often can polyhemoglobin from another species be infused before there is immunological problem.

IMMUNOLOGICAL EFFECTS OF TRANSFUSION IN RATS

NONIMMUNIZED RATS:

1. Homologous Hb PolyHb
   no adverse reactions
2. Heterologous Hb PolyHb
   no adverst reactions

IMMUNIZED RATS:

1. Homologous Hb PolyHb
   no adverse reactions
2. Heterologous Hb PolyHb
   severe anaphylactic reactions:
   bronchospasm, shock, death

Fig. 28. Summary of the effects of transfusion to immunized and nonimmunized rats as described in the previous two figures.

Estep's group [141], infused diaspirin crosslinked human hemoglobin into Rheus monkeys. With 5 ml/kg monthly infusion, there was no production of IgG or IgM to the crosslinked hemoglobin (Figs. 29, 30). Thus, this type of crosslinking did not result in any increase in antigenicity. Furthermore, their results also show that infusion is far less likely to cause immune response than subcutaneous injections of heterologous hemoglobin with Freund's adjuvant. Thus, they showed that five intravenous monthly infusions of human crosslinked hemoglobin into a different species, rhesus monkey, did not result in any antibody production.

Masking of Antigenicity

Microencapsulated hemoglobin and conjugated hemoglobin have been used to mask the antigenicity of hemoglobin. Our study shows that when injected with Freund's adjuvant, liposome-encapsulated hemoglobin from the same species did not result in an increase in antibody titers (Fig. 25) [86]. Encapsulated hemoglobin from another species resulted in a much lower increase in antibody titer

Fig. 29. Antibody levels (Anti–DCLHb IgG) in Rhesus Monkey after multiple intravenous infusion of diaspirin crosslinked hemoglobin (DCLHb). Results from Estep's group. Reprinted with permission from Estep, TN, J Gonder, I Bornstein, S Young, RC Johnson. Biomaterials, Artificial Cells and Immobilization Biotechnology 1992; 20:603–610. Courtesy of Marcel Dekker Inc.

than the same hemoglobin or heterologous polyhemoglobin (Fig. 25) [86]. Encapsulated hemoglobin from another species was also injected subcutaneously in immunizing doses without the use of Freund's adjuvant. Here there was no significant increase in antibody titers (Fig. 25) [86]. In another study PEG-conjugated bovine hemoglobin was infused at 30% blood volume intravenously at weekly intervals for five weeks; there was no significant increases in antibody titers (Fig. 25). When challenged with the final fifth intravenous injection on the fifth week, there was no anaphylactic reaction or fall in blood pressure observed (Fig. 25). In regard to this infusion study, it should be pointed out that crosslinked hemoglobin when infused using the same protocol also did not produce any antibody response (Fig. 29, 30) [141].

Table II. Analysis by ELISA of Anti-DCLHb IgM Levels in Rhesus Monkey Sera After Multiple I.V. Infusions of DCLHb Solutions

	Optical Density (405 nm) Developed in ELISA of Sera [1,2]			
Test Sera Dilution (v:v)	Preinfusion	Number of Prior DCLHb Infusions		
		1	3	5
1:5	0.103 ± .068	0.121 ± .077	0.092 ± .057	0.099 ± .072
1:10	0.083 ± .046	0.097 ± .060	0.066 ± .033	0.076 ± .045
1:50	0.036 ± .023	0.035 ± .021	0.026 ± .013	0.025 ± .012
1:100	0.019 ± .010	0.018 ± .010	0.014 ± .004	0.014 ± .008

1. Values are means of five samples assayed in duplicate ± one standard deviation. All absorbance values were corrected for background.
2. The average optical density of positive control sera (n=2) at a 1:100 dilution was 0.663 ± .057. The average optical density of a second group of negative control sera (n=12) collected from normal rhesus monkeys never previously exposed to DCLHb was 0.016 ± .009 at a dilution of 1:100.

Fig. 30. Antibody levels (Anti–DCLHb IgM) in Rhesus Monkey after multiple intravenous infusion of diaspirin crosslinked hemoglobin (DCLHb). Results from Estep's group. Reprinted with permission from Estep, TN, J Gonder, I Bornstein, S Young, RC Johnson. Biomaterials, Artificial Cells and Immobilization Biotechnology 1992; 20:603–610. Courtesy of Marcel Dekker Inc.

Summary

The use of Freund's adjuvant is a very sensitive test for immunogenicity. It was used to answer the basic questions of effects of crosslinking on immunogenicity. In clinical applications, modified hemoglobins are of course not used with Freund's adjuvant. Indeed, without the use of Freund's adjuvant, subcutaneous and intravenous injection of heterologous hemoglobin is much less immunogenic. For example, five monthly intravenous infusions of human crosslinked hemoglobin into rhesus monkey did not result in any antibody response. Hemoglobin is known to have low antigenicity and requires large repeated doses to produce antibodies. Further studies are required to establish the relationship between the amount infused and the number of intravenous infusions of heterologous hemoglobin. This is important because of the recent promising developments in bo-

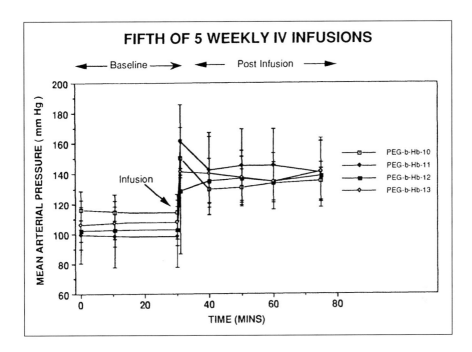

Fig. 31. Rats top loaded with weekly intravenous infusion of 30% blood volume of PEG conjugated bovine hemoglobin (PEG–b–Hb). This is the result of the fifth injection. Reprinted with permission from Chang TMS, Lister C , Nishiya T , Varma R. Biomaterials, Artificial Cells and Immobilization Biotechnology 1992; 20:611–618. Courtesy of Marcel Dekker Inc.

vine polyhemoglobin. Furthermore, transgenic hemoglobin is an potential source of hemoglobin. Here, one may want to know the required degree of antigenic purity for transgenic human hemoglobin. If repeated infusion of large amounts of heterologous hemoglobin is a problem, available immobilization technology like encapsulated hemoglobin and conjugated hemoglobin may be used to mask the antigenicity. Further studies may also be required to study whether the very strict immunization procedure using Freund's adjuvant is required for the assessment of possible hypersensitivity reactions in highly sensitive individuals.

Sterility, Trace Contaminants, Endotoxins and Bridging the Gap Between Safety Studies in Animal and Human

Sterility, Trace Contaminants and Endotoxins

As discussed earlier, it is of the utmost importance to ensure that the modified hemoglobin is free from infective microorganisms, especially hepatitis B and C and HIV. Furthermore, contamination with endotoxin must be carefully screened. The need to remove endotoxin and the interaction of hemoglobin with endotoxin have been emphasized by a number of groups [79, 111, 262, 371, 372, 408, 419]. Potential contamination by chemicals, airborne materials and numerous other possible sources also has to be ruled out. No matter how carefully it is analyzed, it will depend on the available methods. Furthermore, having screened out all potential contaminants by standard analytical methods and having tested the preparation in animal safety studies, we still cannot be sure. This will be discussed in the next section.

Animal Safety Studies Are Not Necessarily Valid for Humans

Response in animals is not always the same as for humans. This is especially true in tests for hypersensitivity, complement activation and immunology. How do we bridge the gap between animal safety studies and use in humans? Complement activation is important in many adverse reactions of humans to modified hemoglobin [150, 317]. Modified hemoglobin may be contaminated with trace amounts of blood group antigens that can form antigen–antibody complexes [21]. This can be detected by complement activation. Other potential materials can also cause complement activation. These include endotoxin, microorganisms, insoluble immune–complexes, chemicals, polymers, organic solvents and others [80, 83, 89, 90, 91, 95].

In Vitro Screening Test Using Human Plasma Before Use in Humans

We have devised an in–vitro test tube screening tests using human plasma or blood (Fig. 32) [80, 83, 89, 90, 91]. The use of human plasma or blood gives the closest response next to actual injection. If we want to be more specific, we can use the plasma of the same patient who is to receive the blood substitute. Many components of human blood or plasma can be used for this in–vitro screening test. If one were to select only one test, perhaps the most useful one would

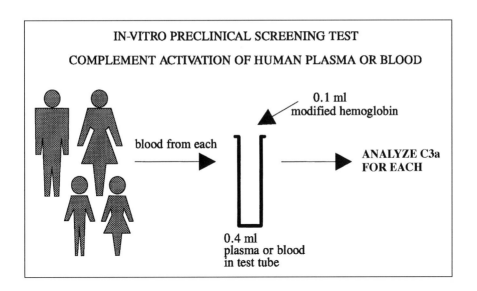

Fig. 32. Use of human plasma or finger–prick blood to test human complement activation response in-vitro to modified hemoglobin. Reprinted with permission from Chang TMS, Lister C. Biomaterials, Artificial Cells and Immobilization Biotechnology 1992; 20:171–180. Courtesy of Marcel Dekker Inc.

be the effect of modified hemoglobin on complement activation (C3a) when added to human plasma or blood (Fig. 32). This simple test consists of adding 0.1 ml of modified hemoglobin to a test tube containing 0.4 ml of human plasma or blood. Then analyze the plasma for complement activation after incubating for one hour (Fig. 32).

Laboratory Procedure Based on Human Plasma

Blood is obtained by clean venous puncture from human volunteers and placed into 5 ml polypropylene (Sastedt) heparinized tubes (10 IU heparin/ml of blood) (Fig. 33). Immediately separate plasma by centrifugation at 5500 G at 2°C for 20 minutes and freeze the plasma in separate portions at –70°C. Do not use serum because coagulation initiates complement activation. EDTA should not be used as an anticoagulant because it interferes with complement activation.

Immediately before use, the plasma sample is thawed. 400 lambdas of the plasma are pipetted into 4 ml sterile polypropylene tubes (Fisher). 100 lambdas of pyrogen free saline (or Ringer's Lactate) for injection is added to the 400

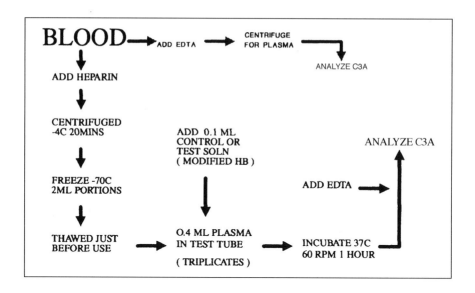

Fig. 33. Summary of procedures in this screening test. Reprinted with permission from Chang TMS, Lister C. Artificial Cells, Blood Substitutes and Immobilization Biotechnology 1992; 20:171–180. With Courtesy of Marcel Dekker Inc.

lambdas of human plasma as control. One hundred lambdas of hemoglobin or modified hemoglobin are added to each of the other tubes containing 400 lambdas of human plasma. The reaction mixtures are incubated at 37°C at 60 rpm for 1 hour in a Lab–Line Orbit Environ Shaker (Fisher Scientific, Montreal, Canada). After 1 hour the reaction is quenched by adding 0.4 ml of this solution into a 2 ml EDTA sterile tube containing 1.6 ml of sterile saline. The samples are immediately stored at –70°C until analyzed.

The analytical kit for human complement C3a is purchased from Amersham Canada. The method of analysis is the same as the instructions in the kit with two minor modifications. Centrifugation is carried out at 10,000 g for 20 minutes. After the final step of inversion, the inside wall of the tubes is carefully blotted with Q–tips.

The exact procedure and precise timing described above are important in obtaining reproducible results. The base line control level of C3a complement activation will vary with the source and procedure of obtaining the plasma. Therefore, a control baseline level must be used for each analysis. Furthermore, all control and test studies should be carried out in triplicates. Much practice is needed to establish this procedure when used for the first time. Reproducibility must be established before this test can be used.

Procedure Based on Blood from Finger Pricks

Instead of using plasma and the need to withdraw blood with a syringe, a simpler procedure involves obtaining blood from finger pricks as follows [90]. Sterile methods were used to prick a finger. Blood was collected in heparinized microhematocrit tubes. The tubes were kept at 4°C, then used immediately as follows. Each blood sample used in testing contains 80 μl of whole blood and 20 μl of saline. Each test solution is added to a blood sample. Test solutions include saline (negative control); Zymosan (positive control) or hemoglobin. This is incubated at 37°C at 60 rpm. After 1 hour of incubation, EDTA solution is added to the sample to stop the reaction. The analysis for C3a is then carried out as described. The test kit is based on ELISA C3a Enzyme immunoassay (Quidel Co, San Diego, CA, U.S.A.).

Areas of Application

This is based on the pathway of complement activation of C3 into C3a (Fig. 34). Factors in modified hemoglobin that can initiate alternate and classical pathways are shown in Figure 35. For example, in research, development or industrial production of blood substitutes, different chemicals, reagents and organic solvents are used. This includes crosslinkers, lipids, solvents, chemicals, polymers and other materials. Some of these can potentially result in complement activation and other reactions in humans. Other potential sources of problems include trace contaminants from ultrafilters, dialysers and chromatography.

Use in Research

Why wait for the completion of research, industrial production and preclinical animal studies (Fig. 36)? Why don't we do this test right at the beginning during the research stage (Fig. 36)? If a new system is found to cause complement activation at this stage, one can avoid a tremendous waste of time and money in further development, industrial production and preclinical animal study [95].

In our ongoing study of hemoglobin nanocapsules, different polymers, lipids, reagents and solvents are used. We therefore analyzed their effects on complement activation of human plasma in–vitro [95]. The results are shown in Fig. 37. One type of L–phosphatidylcholine caused marked increases in complement activation. Another type of L–phosphatidylcholine did not result in marked increases in complement activation. Polymers tested like polylactic acids and ethylcellulose did not result in this degree of complement activation. After repeated washing,

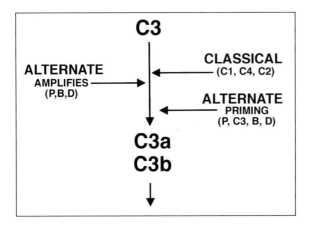

Fig. 34. Alternate pathway and classical pathways in complement activate and conversion of C3 to C3a and C3b. Reprinted with permission from Chang TMS, Lister C. Biomaterials, Artificial Cells and Immobilization Biotechnology 1992; 20:171–180. Courtesy of Marcel Dekker Inc.

CLASSICAL PATHWAY
C1 FIXING
ANTIGEN–ANTIBODY COMPLEXES
FROM IgG OR IgM ANTIBODIES
(IgM BLOOD GROUP ANTIBODIES)

ALTERNATE PATHWAY
- VARIOUS MICROORGANISMS
- ENDOTOXINS
- INSOLUBLE IMMUNE COMPLEXES
  example: from IgA
- VARIOUS CHEMICALS, POLYMERS
- ORGANIC SOLVENTS, ETC.
- OTHERS

Fig. 35. Potential contaminants in modified hemoglobin that might be decided using this screening test. Reprinted with permission from Chang TMS, Lister C. Biomaterials, Artificial Cells and Immobilization Biotechnology 1992; 20:171–180. Courtesy of Marcel Dekker Inc.

- **RESEARCH STAGE:**
  POLYMERS, LIPIDS, XLINKERS, CHEMICALS, SOLVENTS

- **ANIMAL SAFETY STUDIES**
  HUMAN RESPONSE NOT ALWAYS = ANIMAL

- **INDUSTRIAL PRODUCTIONS**
  BATCH VARIATIONS, CONTAMINATIONS, ETC.

- **CLINICAL TRIALS & USES**
  PRECLINICAL TRIAL FINAL SAFETY TEST
  ANALYSIS OF POPULATION RESPONSE

Fig. 36. Potential uses of screening test in the different stages of research, development and clinical trial of modified hemoglobin.

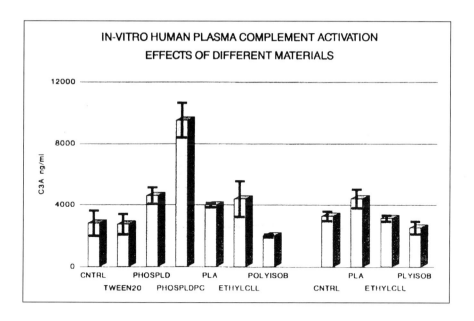

Fig. 37. Screening test used in the research stage to analyze the different chemicals and reagents used in the preparation of biodegradable polymeric hemoglobin nanocapsules. Reprinted with permission from Chang TMS, Lister C. Artificial Cells, Blood Substitutes and Immobilization Biotechnology, an International Journal 1994; 22: 159–170. Courtesy of Marcel Dekker Inc.

there was no longer any significant complement activation (Fig. 37). Another polymer, isobutyl 2–cyanoacrylate, resulted in less C3a level than the control. The reason for this is that the polymer does not cause complement activation while it adsorbs C3a. The emulsifying agent Tween 20 also did not result in complement activation. Other groups are starting to use this in research [410].

Use in Industrial Production

We have also used this preclinical test to help others in industrial production [91, 95]. Thus in an early industrial scale–up of polyhemoglobin, this in–vitro screening test showed that certain batches caused complement activation. By using this test further, it shows that this is the result of the use of new ultrafiltrators. Re–used or washed ultrafiltrators did not cause complement activation (Fig. 38). Without this test, some batches could result in adverse effects of "unknown causes" in humans. Chromatography, ultrafiltrators, dialysis membranes and other separation systems are used extensively in the preparation of different types of blood substitutes. It is therefore important to screen for the possibility of trace contaminants that could cause complement activation. In the same way, different chemical agents and different reactants used in industrial production could be similarly tested.

Correlation of In Vitro Complement Activation to Clinical Symptoms

What are the clinical implications of C3a levels in the above in–vitro complement activation screening test? Until actual clinical data is available, one cannot conclusively establish this. On the other hand, there are clinical data in the use of different types of hemodialysis membranes and their relationships to complement activation and clinical symptoms (Deane and Wineman 1988). In these patients, clinical symptoms appeared when C3a was substantially elevated. Increasing levels resulted in increasing severity in the symptoms. Symptoms include myalgia, chest tightness, fever, chill and others.

C3a is a smaller molecule (M.W. 9,000) than C3 (M.W. 180,000). Therefore once it is formed, C3a equilibrates rapidly across the capillaries. Thus, in measuring C3a in patients one has to do this within very short intervals to catch the peak rise in C3a. The peak is reached in the first 15 minutes. After this, it declines rapidly to normal in 60 min. In following complement activation in patients receiving blood substitute, one has to catch the peak use. In the in–vitro study, the C3a would not escape from the test tube, therefore, the maximal level of C3a would be available.

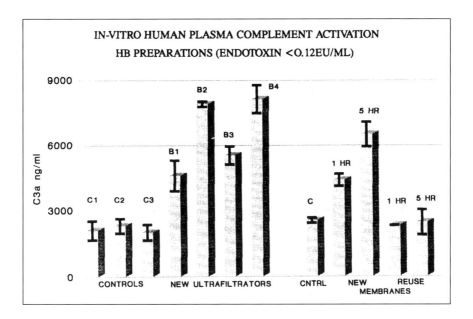

Fig. 38. Screening test used in the development and production stage to find out and rule out why certain batches of modified hemoglobin cause potential problems. In this case, it was shown that new ultrafiltration membranes releases contaminants into modified hemoglobin. These batches caused complement activation. Careful washing of the ultrafiltrator membranes (used) has solved this problem. Reprinted with permission form: Chang TMS, Lister C. Artificial Cells, Blood Substitutes and Immobilization Biotechnology, an International Journal 1994; 22:159–170. Courtesy of Marcel Dekker Inc.

Clinical Trials and Use in Humans

This in–vitro test may be useful in large scale screening for human response [95]. For instance, it could be used to study variations in production batches. It could also be used to study individual variations. Furthermore, it could also be used to analyze the response of different human populations, especially with different disease conditions. It is important to note that all these could be done without ever introducing any blood substitute into humans.

Summary

This is a very simple in-vitro test based on human plasma or blood. For example, this screening test can detect several problems related to potential hypersensitivity, and anaphylactic reactions, effects due to antibody–antigen complexes and others. These potential problems may be related to contamination with trace blood group antigens, polymeric extracts, organic solvents, emulsifiers and others. This is useful for preclinical trial studies and for screening before clinical use. This is also useful for screening batches of modified hemoglobin blood substitutes for industrial production to rule out potential problems. It is also useful in research and development. What is potentially very exciting is that this approach may be the basis of a large–scale "clinical trial" in a large number of patients without infusing the product [95]. By doing this, we can analyze the percentage of patients who may have adverse reactions without having to introduce the product into patients.

Nitric Oxide, Molecular Sizes, Vasoactivities and Gastrointestinal Effects

Vasoactivities

This is an important area of investigation. The earlier work of Vogel and Valeri on vasoactivities, using the rabbit heart coronary artery model, has contributed extensively to this area [442–444]. This is an important area of research for modified hemoglobin since hemoglobin preparations may cause vasoconstriction resulting in decreased regional perfusion and increased blood pressure. More recently, interactions between hemoglobin and nitric oxide have become very important [115, 225, 246, 247, 364, 432, 433].

Endothelial Cells and Nitric Oxide

Hemoglobin has high affinity for nitric oxide and may cause vasoconstriction in the following way. Blood vessels are very schematically represented in Figure 39. Endothelial cells line the lumen of the vessel (Fig. 39). Endothelial cells, through a number of steps, release nitric oxide into the interstitial space of the vessel wall. Nitric oxide causes the smooth muscle of the vessel wall to relax. This way the amount of nitric oxide released can control the vasoactivity of the vessel wall. Changes that lower nitric oxide result in vasoconstriction. Increasing nitric oxide results in vasodilatation. Nitric oxide also acts on the nerve plexus and other sites in the body.

The intercellular junctions of the endothelial cell layer allow tetrameric hemoglobin to cross from the circulating blood (Fig. 40). Since hemoglobin binds nitric oxide, this tetrameric hemoglobin on leaving the circulation acts as a sink

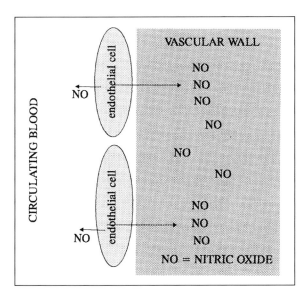

Fig. 39. Endothelial cells of the blood vessels release nitric oxide (NO). NO causes the smooth muscle of the vessel wall to relax. When NO is removed, blood vessels constrict (vasoconstriction). Thus, NO plays a role in the control of the vascular tone. Reprinted with permission from Chang TMS. Artificial Cells, Blood Substitutes and Immobilization Biotechnology, an International Journal 1997; 25:1–24. Courtesy of Marcel Dekker Inc.

in removing nitric oxide resulting in vasoconstriction. Murray et al [310] reported that esophageal spasm at higher doses of some types of modified hemoglobin could also be due to the removal of nitric oxide by tetrameric hemoglobin in this way.

Managing the Effect of Tetrameric Hemoglobin on Nitric Oxide (Fig. 41).

Those who are using intramolecularly modified tetrameric hemoglobin are using this property as a therapeutic application for hypotensive shock [351]. Blood substitutes prepared from polyhemoglobin, encapsulated hemoglobin and conjugated hemoglobin, depending on the preparation, contain a variable amount of tetrameric hemoglobin. This tetrameric hemoglobin can also cross the intercellular junction of the endothelial cells. Those who are not using this effect are preparing their blood substitutes to avoid this [174–176]. One way is to add

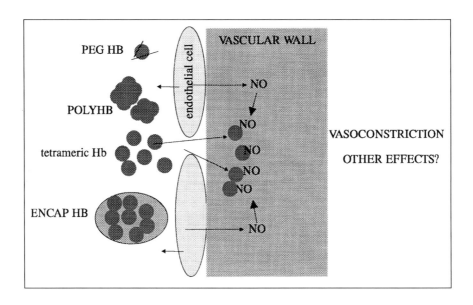

Fig. 40. Different types of modified hemoglobin (Hb). Tetrameric Hb being smaller can move across the intercellular junction of the endothelial cells. This Hb binds NO and therefore lowers the NO concentration. This results in vasoconstriction. Other types of modified hemoglobin may contain a varying amount of tetrameric hemoglobin which can also act similarly. Reprinted with permission from Chang TMS. Artificial Cells, Blood Substitutes and Immobilization Biotechnology, an International Journal 1997; 25:1–24. Courtesy of Marcel Dekker Inc.

pharmaceutical agents to counteract the effects. Another group is using recombinant technology to change one amino acid to block the nitric oxide adsorption site of the resulting recombinant hemoglobin. One common way is to remove as much tetrameric hemoglobin as possible from the preparation (Fig. 42). Thus, Gould et al have emphasized the need for this and have prepared polyhemoglobin with <1% tetrameric hemoglobin for their clinical trials [174–176]. This allows them to infuse 3000 ml of the preparation with no reported side effects. Shorr et al recently also reported on the removal of terameric hemoglobin from PEG–conjugated hemoglobin to lessen esophageal effects [397].

What happens if one uses polyhemoglobin, conjugated hemoglobin or encapsulated hemoglobin containing no tetrameric hemoglobin? Theoretically, there will not be a sink of tetrameric hemoglobin to remove nitric oxide and all the effects associated with this should be eliminated (Fig. 42). However, nitric oxide can also enter the blood stream from the lumen side of the endothelial cells.

WHAT TO DO WITH NITRIC OXIDE EFFECTS?

REMOVE TETRAMERIC HB

MAKE USE OF THIS PROPERTY

RECOMBINANT MODIFICATION OF HB

MEDICATIONS AGAINST EFFECTS

OTHER NEW APPROACHES

Fig. 41. Different approaches in managing the effects of tetrameric Hb on nitric oxide (NO). Reprinted with permission from Chang TMS. Artificial Cells, Blood Substitutes and Immobilization Biotechnology, an International Journal 1997; 25:1–24. Courtesy of Marcel Dekker Inc.

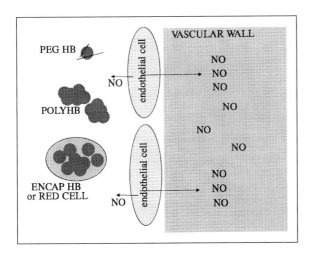

Fig. 42. Removing all tetrameric hemoglobin from the modified Hb preparation should prevent the Hb from crossing the intercellular junction and thus prevent vasoconstriction. Reprinted with permission from Chang TMS. Artificial Cells, Blood Substitutes and Immobilization Biotechnology, an International Journal 1997; 25:1–24. Courtesy of Marcel Dekker Inc.

How do hemoglobin blood substitutes behave toward this source of nitric oxide? Can one expect any effects when a large amount of blood substitute is infused? In the same way, how does hemoglobin in red blood cells behave toward this source of nitric oxide? This will be discussed in the chapter on future perspectives of blood substitutes.

Chapter 6

Blood Substitutes: Principles, Methods, Products and Clinical Trials,
by Thomas Ming Swi Chang. © 1997 Karger Landes Systems.

••••••••••••••••••••••••••

What Are the Present Modified Hemoglobin Products Being Tested in Clinical Trials in Patients?

The blood substitute products now in clinical trials can be divided into the different classes of modified hemoglobin as follows:

1. Polyhemoglobin from intermolecular crosslinking of hemoglobin:
   Pyridoxalated human polyhemoglobin) – Northfield Laboratory
   Pyridoxalated bovine polyhemoglobin – Biopure Corp.
   o–Raffinose human polyhemoglobin – Hemosol Ltd.
2. Intramolecularly crosslinked or recombinant modified tetrameric hemoglobin:
   Diaspirin crosslinked hemoglobin DCLHb – Baxter Healthcare
   Recombinant human hemoglobin – Somatogen
3. Conjugated hemoglobin from hemoglobin crosslinked to soluble polymers:
   Pyridoxalated polyoxyethylene conjugated human hemoglobin –
   Ajinomoto and Apex Bioscience
   Polyethylene glycol conjugated bovine hemoglobin – Enzon, Inc.

Pyridoxalated Human Polyhemoglobin (Polyheme™) – Northfield Laboratory

Northfield Laboratories Inc. has the longest experience in the clinical assessments of modified hemoglobin [171–176, 304–305, 383–386]. The blood substitute here is Polyheme™ based on pyridoxalated human polyhemoglobin. As discussed earlier, this is the most studied type of modified hemoglobin. Gould will discuss this in a later chapter.

General Characteristics

Northfield has placed emphasis on removing tetrameric hemoglobin from its preparation. This is to prevent the extravasation of these tetramers from the capillary to bind nitric oxide as discussed earlier. Northfield reported that this will avoid problems related to vasoconstriction, gastrointestinal effects and other problems. To do this, Northfield used a two–step process to create a tetramer–free form of polymerized hemoglobin. The first step is to polymerize hemoglobin resulting in an array of different size molecules. The next step is the removal of all unreacted tetramer. The final preparation has a tetramer of less than 1% that is less than those in any other products. The key characteristics for the Northfield preparation being used in clinical trials as described recently [173–176] are: hemoglobin concentration 10 g/dl; P_{50} 28–30 mmHg; methemoglobin <3%; tetramer <1%. One unit of 500 ml therefore contains 50 gm of hemoglobin.

Results of Phase I Clinical Trials in Human

Gould et al [171] in 1993 reported on their phase I clinical trials using pyridoxalated polyhemoglobin. Healthy adult males were screened to serve as the experimental subjects. The volunteers were hydrated intravenously with Ringer's lactate and orally with water. After a baseline period of hydration, polyhemoglobin was infused. Hemodynamics, hematologic and biochemical profiles and renal functions were assessed hourly before, during and after infusion. Infusion of up to one unit (50 gm hemoglobin) of this preparation into healthy volunteers did not produce any adverse effects. Hemodynamics did not change. There was no gastrointestinal distress. Renal function and other organ functions were normal throughout the study period. Furthermore, polyhemoglobin infusion did not cause vasoconstriction in human volunteers.

Results of Phase II and III Clinical Trials and Efficacy Studies in Human

The above safety results allowed Northfield to go to phase II efficacy clinical trials [173–176]. Having established the safety of crosslinked hemoglobin in humans, attention is now turned to the study of efficacy in human clinical trials. They have recently summarized the results of their phase II clinical trial [176]. This is for treating the acute blood loss that occurs following trauma or during surgery.

In the first part of their clinical trials, the protocol involves infusing up to three units (150 gm). When the clinical investigators decided that transfusion is indicated, the patient would receive 1, 2 or 3 units of the preparation in place of red blood cells. Thirty patients received the following amounts: 14 patients have received 1 unit; 2 patients received 2 units and 14 patients received 3 units. They have not observed any safety concerns in these patients. They are addressing efficacy by calculating the total concentrations of hemoglobin in the presence of red blood cells and polyhemoglobin [176]:

total [Hb] = RBC [Hb] + Polyhemoglobin [Hb] (measured as plasma hemoglobin)

The results show that in the 30 patients, infusion of 1, 2 or 3 units of the polyhemoglobin can replace the loss of red blood cells. Thus, the values (g/dl) for red blood cells (and total hemoglobin) are as follows [176]:

Preinfusion:	rbc: 9.7 ± 2.3	total hemoglobin: 9.7 ± 2.3
After 1 unit	rbc: 8.6 ± 1.8	total hemoglobin: 9.5 ± 1.6
After 2 units	rbc: 6.4 ± 1.2	total hemoglobin: 8.6 ± 0.9
After 3 units	rbc: 5.6 ± 1.1	total hemoglobin: 8.6 ± 0.4

In addition, 12 of the 30 patients who would have needed to receive allogeneic transfusion during the first 24 hours did not need this after the polyhemoglobin infusion. They increased infusion to 6 units (3,000 ml) of polyhemoglobin and the results continue to show the safety and efficacy of this preparation [176]. Clinical trials are being continued at increased doses (up to 10 units) and in a randomized, controlled study for comparing safety and efficacy with that of allogeneic blood. The significance of this clinical trial approach is that it reduces allogeneic transfusion during the treatment of hemorrhage. Northfield has received F.D.A. approval to go on with phase III clinical trials in surgical patients.

Pyridoxalated Bovine Polyhemoglobin – (Hemopure™) Biopure Corp.

General Characteristics

Pearce and Gawryl from Biopure Corp. will discuss their blood substitute in a separate chapter in the second volume of this book. Briefly, they have used the most studied method of glutaraldehyde polymerized hemoglobin. However, instead of human hemoglobin, they use bovine hemoglobin. Bovine hemoglobin is easily available in large amounts. Furthermore, unlike human hemoglobin, even without 2,3–DPG or its analog pyridoxal–phosphate, bovine hemoglobin has a much higher P_{50} than human hemoglobin. There were initial worries of the immunogenicity of polyhemoglobin prepared from bovine hemoglobin. As discussed earlier, animal studies showed that hemoglobin from a different species when infused for the first time did not cause immunological problems [75, 86]. There were immunological reactions only when this is infused into animals immunized beforehand [75, 86]. In nonimmunized animals, as discussed earlier, repeated infusion of small doses did not give rise to antibody formation.

Phase I and II Clinical Trials in Humans

Hughes et al [206] in 1993 reported their phase I clinical studies. More recently, Jacobs et al [222] updated their phase II clinical studies on polymerized bovine hemoglobin (HBOC–201). This is currently under clinical development for trauma, and perioperative surgical uses and for the treatment of sickle cell crisis. HBOC–201 is composed of glutaraldehyde polymerized bovine hemoglobin formulated in a balanced electrolyte solution at a concentration of 13 g/dl. This is stable at room temperature for greater than one year. The viscosity is 1.3 centipoise and P_{50} is 38 mmHg. Clinical pharmacology of this preparation shows that it is effective in maintaining exercise capacity after acute anemia. This may be related to an enhancement of oxygen on–loading and off–loading as shown by increase in pulmonary diffusing capacity in normal volunteers and in tissue oxygen studies in a canine model. Up to 166 g have been given to more than 200 human subjects. They reported that safety has been shown during these studies and in phase II clinical trials in progress.

o–Raffinose Human Polyhemoglobin (Hemolink™) – Hemosol Ltd.

General Characteristics

This group will be describing their research in a later chapter by Adamson and Moore in the second volume of this book. A brief preview of this chapter is as below:

Briefly, Hemosol has developed and extended the use of a polyaldehyde, o–raffinose, prepared from the oxidation of raffinose originated by Hsia [200, 201]. o–Raffinose can crosslink hemoglobin intermolecularly and intramolecularly. Intermolecular crosslinking, as in glutaraldehyde crosslinking of surface amino groups, resulted in the polymerization of hemoglobin molecules. The intramolecular crosslinking between amino groups within the 2,3–DPG binding pocket under specific reaction conditions contributes to high P_{50}. Pliura et al from this group have developed a self–displacement chromatographic process that resulted in highly purified HbA_{0} [345]. Hemolink is based on the use of a modified o–raffinose method to crosslink HbA_{0}.

The product is o–raffinose human polyhemoglobin at a concentration of 10 g/dl in lactated Ringer's solution. Methemoglobin is <10%, P_{50} is 34 mmHg, oncotic pressure is 24 mmHg, polyhemoglobin is about 63 +/– 12%. This was tested in preclinical studies in animals. The models used were top–load, model, hypovolemic model and isovolemic exchange model. Renal function studies were carried out in details. Other studies included immunology, complement activation, vasoactivity, circulation half–time, biodistribution and others.

Phase I and II Clinical Trial in Human

Based on the above preclinical result, they received approval to carry out phase I clinical trial to evaluate the safety. As discussed in a later chapter by Adamson and Moore, they reported no clinically significant changes, except gastrointestinal discomfort at the highest dose – similar to those reported in most of the other types of blood substitutes. They have now received approval to carry out phase II clinical trial in surgical patients.

Diaspirin Crosslink Hemoglobin (DCLHb) HemAssist™ – Baxter Healthcare

General Characteristics

Details are available in a later chapter by Nelson in the second volume of this book. Very briefly, unlike polyhemoglobin described above it is an intramolecularly crosslinked tetramer. Walder et al in 1979 used bis(3,5– dibromosalicyl) fumarate (DBBF) to intramolecularly crosslink the two α subunits of the hemoglobin molecule [446]. This prevents dimer formation and improves $P_{50.}$ The Baxter group under Estep has developed several improvements including heat viral–inactivation, large scale production and other extensions [39, 141, 147]. Each blood substitute is developed to fulfill specific function. For instance, the pyridoxalated polyhemoglobin developed by Northfield as described above, emphasized the removal of all tetrameric hemoglobin to avoid its pressor effects. For DCLHb, the application is focused on its ability to raise blood pressure. Research shows that the vasoactivity of DCLHb is only partly due to NO binding, because there is also up–regulation of endothelin [188] and perhaps also adrenergic mechanisms. Gulati et al have carried out detailed animal studies on the hemodynamics [16, 187, 188, 397]. Animal studies by the same group show improved tissue perfusion in hemorrhagic shock when only a small amount of DCLHb was administered [255].

Phase I Clinical Trial in Human

Przybelski et al in 1993 reported their phase I clinical trial using diaspirin crosslinked hemoglobin (DCLHb) [350]. They used a randomized, placebo controlled, double blind protocol of crossover design. Twenty–four participants received one of 25, 50 or 100 mg/Kg of the 10 g/dl DCLHb or an equal volume of lactated Ringer's over 30 minutes. Each is followed in five days by infusion of the alternate solution. A dose–related increase in mean arterial pressure was observed after DCLHb infusion with an associated decrease in heart rate. No evidence of vasoconstriction was observed. The LDH–isoenzyme fraction was increased in the DCLHb recipients, but the total LDH was not elevated. There were no myalgia, temperature elevations or associated symptoms. The infusion was well tolerated.

Phase II and III Clinical Trials in Patients

Przybelski et al in 1996 [351] reviewed and updated their clinical studies. Their initial clinical studies have been followed by a multinational efficacy study of its use in cardiac surgery patients starting in mid 1995. The first safety studies in patients examined the safety of this preparation in 130 hemorrhagic shock patients at ten sites in the U.S. and Europe. They reported that DCLHb was shown safe at the doses tested. They therefore started efficacy studies in trauma patients. They also evaluated the use of the pressor effect of DCLHb in a hemodialysis patient study and reported that there was greater blood pressure stability with DCLHb.

The pressor effect was studied further in three surgical studies involving 160 patients. Here patients received DCLHb as prophylaxis against hemodynamic instability during aortic repair [161], orthopedic or major abdominal surgeries. In a critically ill/ICU study, 14 "sepsis syndrome" patients with low systemic vascular resistance despite maximum standard therapy were given DCLHb. They reported a rapid and significant vasopressor response that allowed for a 15–100% reduction in standard vasopressor drug requirements, with a significant decrease in mean APACHE II scores 24 hours after treatment [359]. They also reported that in addition to the pressor effects, there is volume expansion. They are now using DCLHb as a more classical blood substitute, in the perioperative period after cardiopulmonary bypasses at eight European sites. They are also starting efficacy studies in trauma patients.

Recombinant Human Hemoglobin – Somatogen

General Characteristics

This group will contribute a chapter in the second volume of this book. This is based on the basic finding by Hoffman et al that human hemoglobin can be expressed in *E. coli* with the introduction of the human genes [196]. Somatogen's groups, Looker et al, Shoemaker et al, extended and developed this approach resulting in a human hemoglobin derivative, di–alpha–hemoglobin [264, 265, 395, 396]. In this approach, the two α subunits of the hemoglobin molecule are fused, thus preventing the breakdown of the tetrameric hemoglobin into dimers. In addition, they can modify the hemoglobin by further mutation to have a P_{50} better than that of human hemoglobin. Somatogen has developed large–scale production with a high yield of products, a high level of gene expression and purification from *E. coli* contaminants.

Phase I and II Clinical Trial

Shoemaker et al in 1993 [396] reported their initial experience in their phase I safety clinical trial using recombinant human hemoglobin. The subjects were given rHb1.1 or normal serum albumin intravenously blinded to the subject. The maximum dose was 11 g given in 0.8 hours. There was no evidence of major organ system toxicity. Renal function was normal with no evidence of renal clearance of rHb1.1. There was no evidence of hypertension.

In a review by Winslow [497] and presentation by Caspirin [45] in a recent international symposium, the clinical trial results have been reviewed. As with several other modified hemoglobin blood substitutes, they first reported the observation of gastrointestinal side effects with higher doses. This can be avoided by giving the proper medications. Their phase I safety clinical trials were successful and Somatogen started phase II clinical trials in 1994. Their nuclear magnetic resonance (NMR) studies show that this preparation delivers oxygen to muscle. By 1995, they have infused up to 100 grams into each patient with no significant adverse effects. In 1996, they initiated clinical trials in patients with end stage renal failure. They have also initiated clinical trials in patients with refractory anemia in conjunction with exogenous erythropoietin. It seems to act synergistically with erythropoietin to stimulate bone marrow production of red blood cells. They are carrying out a multicenter clinical trial in intraoperative surgical blood loss. Doses given are between 25–100 grams for each patient. A larger multicenter clinical trial is ongoing.

Pyridoxalated Hemoglobin Polyoxyethylene – Ajinomoto and Apex Bioscience

General Characteristics

This is based on soluble conjugated hemoglobin as discussed earlier. In this case the soluble polymer used is polyoxyethylene, a derivative of polyethylene glycol. The hemoglobin used by this group is human hemoglobin. Pyridoxal phosphate is applied as an analog for 2,3–DPG. The resulting product is pyridoxalated hemoglobin polyoxyethylene (PHP). The solution has equal amounts of PHP and maltose to increase stability and prevent methemoglobin formation. Iwashita et al from Ajinomoto has carried out development of this approach for many years [220, 221]. This included extensive animal studies with their collaborators including Agishi et al [2, 4], Matsumura et al [280–282], Nose [293], Sekiguchi et al [387] and others.

Phase I Clinical Trials

This has been summarized in a recent review [457]. This has been approved for phase I clinical trials by the FDA. They joined efforts with Apex Bioscience, U.S.A. Apex has carried out phase I clinical trial using doses of up to 7 g/patient without reporting any significant toxicity. They are concentrating on septic shock as their primary application. This is based on their view that in septic shock, hypotension and reduced tissue perfusion are the result of an excess of nitric oxide. They are using PHP to remove this excess nitric oxide. This is another example of the different uses for blood substitutes.

Polyethylene Glycol (PEG) Conjugated Bovine Hemoglobin – Enzon, Inc.

General Characteristics

Enzon Inc. has already developed PEG–enzymes for applications in enzyme therapy. It has completed clinical trials in specific applications for enzyme therapy. They have followed this up with PEG–hemoglobin. Here they use the soluble polymer polyethylene glycol (PEG) to crosslink to bovine hemoglobin. Their group under Shorr has carried out detailed preclinical studies and reported on the safety of this preparation [314–351, 397, 398]. As reported earlier, PEG when crosslinked to bovine hemoglobin, seems to "cover" the antigenic sites and made it nonantigenic even after five weekly infusions in animal studies [86].

Clinical Trial

Shorr et al in 1996 [397] reported their phase 1B safety evaluation of PEG hemoglobin. They tested this as an adjuvant to radiation therapy in human cancer patients. This is based on their earlier observation that PEG–bovine hemoglobin oxygenates hypoxic tumor tissue and dramatically increases sensitivity to radiation therapy in laboratory models and veterinary patients. Safety evaluation (phase 1a) in healthy human male volunteers has shown that this is well tolerated and remains in circulation long enough to be consistent with weekly dosing and current fractionated radiation therapy practice. They have therefore initiated phase 1b clinical trials in human cancer patients. They also reported on the removal of tetrameric hemoglobin from their PEG–hemoglobin to reduce gastrointestinal effects.

Research is being carried out to use the high oncotic pressure in acute normovolemic hemodilution and in hypovolemic shock [457].

Blood Substitutes: Principles, Methods, Products and Clinical Trials, by Thomas Ming Swi Chang. © 1997 Karger Landes Systems.

••••••••••••••••••••••••••

Blood Substitutes Based on Perfluorochemicals

In the second volume of this book Professor Reiss will discuss the chemistry and developments and Dr. P. Keipert's chapter will emphasize preclinical and clinical studies. What follows is a brief overview of this large area.

What Are Perfluorochemicals?

Of the synthetic organic material, silicone and fluorocarbon are known for their ability to carry oxygen. Figure 43 summarizes the earlier development of this approach. Thus in the 1960s Clark and Gollan [109] showed that mice immersed in oxygenated silicone oil or liquid fluorocarbon could breathe in the liquid. In the same year Chang [51] showed that artificial cells formed from a hybrid of silicone rubber and hemolysate were very efficient in carrying and releasing oxygen. However, these solid elastic silicone rubber artificial cells were removed rapidly from the circulation. Sloviter and Kamimoto in 1967 [403] showed that perfusion using finely emulsified fluorocarbon could maintain rat brain function for several hours. Geyer, Monroe and Taylor [164] in 1968 showed that finely emulsified fluorocarbon could replace essentially all the blood of rats with the rats surviving and recovering. This exciting demonstration did not immediately lead to clinical application. This is because F–Tributylamine available at that time had a long retention time with a $T_{1/2}$ of more than 800 days in the RES. Naito and Yokoyama [312] extended and developed this to produce in 1976 Fluosol–DA 20 suitable for clinical testing.

FLUOROCHEMICALS

(First reports)

1966(Clark)   FC LIQUID & EMULSION

1967(Sloviter)   FC EMULSION PERFUSION

1968(Geyer)   FC EMULSION REPLACED BLOOD

1975(Naito&Yokoyama) FLUOSOL CLINICAL TRIALS

Fig. 43. History in the initial studies of perfluorochemicals as blood substitutes

Perfluorocarbons: Fluosol–DA (20%)

Fluosol–DA is a 20% (w/v) mixture of seven parts of perfluorodecalin and three parts perfluorotripropylamine, with 2.7% pluronic F–68 as an emulsifier and 0.4% of egg yolk phospholipids to form a membrane coating on the emulsion. The average particle size of the emulsion is 0.118 micron. Unlike the earlier fluorocarbon that has tissue retention $T_{1/2}$ of more than 800 days, perfluorodecalin has a $T_{1/2}$ of 7.2 days. Unfortunately perfluorodecalin cannot be used to form stable emulsion and perfluorotripropylamine with a $T_{1/2}$ of 64.7 days has to be combined to form a stable emulsion. The much shorter retention time of the fluosol–DA 20 allows its use for clinical trials and testing. Because of the viscosity of the fluorocarbon emulsion at high concentrations, the maximum amount used is only 20%. Patients have to breathe 70–90% oxygen for the small amount of fluosol–DA to carry enough oxygen. Another reason for needing high oxygen is that oxygen can dissolve in fluorocarbon but cannot bind to hemoglobin. The oxygen carrying capacity is therefore much lower than that of hemoglobin. CO_2 also dissolves in the fluorocarbon that carries it to the lung for excretion.

Results of Clinical Trials Using Fluosol–DA (20%)

Overall Results

Initial clinical testing was carried out in decerebrated human and normal human volunteers by Naito et al in 1978 [312]. After this, larger scale clinical trials were carried out. Mitsuno and Ohyanagi in 1985 reviewed the Japanese experience on 401 patients who received fluosol–DA (20%) [294]. In 270 patients fluosol–DA was given for replacement of blood loss due to hemorrhage. Fluosol–DA was given to 107 patients to improve impaired cerebral circulation, and in others for treating carbon monoxide intoxication, severe autoimmune hemolytic anemia and conditions with impaired blood flows. A maximum of 25 ml/kg body weight of fluosol–DA was usually given. In hemorrhagic patients with blood pressures less than 100 mmHg and pulse rates greater than 100/min, infusion of fluosol–DA (20%) improved blood pressures and returned pulse rates to normal. These patients also maintained their hemodynamic status and maintained or improved their blood gas status. In patients with central venous pressure (CVP) lower than 5 cm water, the CVP returned to normal.

There were no hypotensive episodes and there were no changes in ECG, respiration, body temperature, or bone marrow function. There were slight decreases in platelet counts and some prolongation of bleeding time in some patients but no hemorrhagic complications were observed. The half–life of fluosol–DA in the circulation is related to the dose given. For 10, 20 and 30 ml/kg body weight the $T_{1/2}$ was 7.5 hours, 14.5 hours and 22 hours respectively.

With the high inspired oxygen concentration, the arterial blood oxygen tension was maintained above 300 mmHg whereas the venous oxygen tension was always less than 50 mmHg. From this, it was calculated that of the total oxygen consumed by the tissue, 17% was provided by the fluorocarbon, an amount equal to that provided by the plasma phase. Larger doses of fluosol–DA given to a patient with hemoglobin level of 5.6 g/dl gave the following results. The arterial oxygen content was 6.7 ml/dl in the hemoglobin phase, 1.39 ml/dl in the fluorocarbon phase and 0.96 ml/dl in the plasma phase. The investigators felt that this can result in effective increases in oxygen carriage. This suggestion has been supported by animal studies by other groups. However, Moss' group reported that in their control clinical trial, they did not find significantly useful contributions of oxygen by fluosol–DA [305]. Complement activation was the major clinical problem. Adverse effects were observed in some patients due to complement activation caused by the surfactant used in fluosol. The FDA approved the clinical use of fluosol–DA only for coronary artery balloon angioplasties in 1989.

Modern Perfluorochemicals

This will be discussed in detail in two later chapters. Two new types of preparations have been developed. One type is based on perfluoroctyl bromide ($C_8F_{17}Br$) and perfluorodichoroctane ($C_8F_{16}Cl_2$). Both types permit the use of higher concentrations of PFC. Oxygent™ is from Alliance Pharmaceutical Corp., San Diego is prepared from perfluoroctyl bromide ($C_8F_{17}Br$) with egg yolk lecithin as the surfactant. The use of egg yolk lecithin instead of surfactant has solved the problem of complement activation [143, 241, 242]. Another approach, Oxyfluor from HemoGen, St. Louis is based on the use of perfluoro–dichoroctane ($C_8F_{16}Cl_2$) with triglyceride and egg yolk lecithin [168].

Perfluoroctylbromide (Oxygent™): Clinical Trials

Safety Studies in Human

Safety clinical trials [141, 143, 242] using Oxygent™ show that with increasing dosages there were no adverse effects until 1.8 g/kg. At this highest dose, there is a transient mild febrile response (38–39°C) starting about four hours after injection and gone by 14 hours. At this dose, there is also a decrease in platelet, returning to normal by day seven. These effects were not observed at lower doses. Similar symptoms of fever, chills and nausea were also reported when using high doses of Oxyfluor [168]. There is also a similar reduction in platelet count at the high dose level. They observed that doses up to 1.8 g PFC/kg could be given without side effects.

Efficacy Studies in Human

Oxygent™ is being used in phase II clinical trials in surgical patients breathing 100% oxygen by Spence's group [404, 445] and others. The use of 0.9 g/kg of Oxygent™ seems able to avoid the need for 1–2 units of blood [245]. Their present emphasis is therefore to study the use of PFC in surgery to offset the need for this amount of blood during surgery. This is to be combined with autologous blood.

Potential Areas of Applications

There are several other potential applications for perfluorochemicals as discussed by Bolin et al [29], Jamieson and Greenwalt [425], Mitsuno and Naito [223], Geyer [294], Peerless et al [338], Tremper [425], Reiss [361] and others. At present, these will be limited to the dosage of less than 1.8 g/kg as discussed above. For example, in thrombosis or embolism, the small PFC particles and the increased oxygen pressure may help the affected tissue. Use in patients who because of religious belief cannot use human blood cells is an important and obvious area. Other applications not related to its use as blood substitutes include studies for use in cancer chemotherapy for increasing oxygen tension of tumor tissues.

Present Status and Future Perspectives

The biggest advantage of perfluorochemicals is that they are synthetic materials that can be chemically produced in large amounts without having to depend on donor blood or other biological sources. Much has been done in the last 10 years to improve this approach. Changing the surfactant has solved the earlier problem of complement activation. Higher concentrations of the new perfluorochemicals can now be used to increase oxygen carrying capacity. Right now this is limited by the dosage of 0.9 g/Kg for human use. This low dosage is partly because of side effects observed in humans at a dosage of 1.8 g/Kg. Here, the patients still have to breathe 100% oxygen.

With further research and development, the problem related to side effects at a higher dosage is likely to be resolved. If this can be resolved then the highest dosage will only be limited by the dosage that would not cause significant suppression of the reticuloendothelial system. In this regard, ever improving perfluorochemicals with decreasing residual time in the reticuloendothelial system are being made available. It is likely that further improvements in perfluorochemicals may also lead to further improvements in oxygen carriage thus further reducing the oxygen required for breathing.

With the rapid increase in progress, what is not possible now may change very quickly with the further improvements of new perfluorochemicals. One should not fall into the same trap we had in the 1960s when research on crosslinked hemoglobin and microencapsulated hemoglobin was not carried out. It was thought then that perfluorochemicals and stroma–free hemoglobin were more promising. Change in new technologies had shown otherwise, but at the expense of unnecessary delay of making these approaches available for routine clinical applications. There is therefore every reason to continue with research and development for the very promising approach of perfluorochemicals. At the very least, perfluorochemicals would be an important blood substitute for patients who cannot use human blood substitutes because of religious belief.

Chapter 8

Blood Substitutes: Principles, Methods, Products and Clinical Trials,
by Thomas Ming Swi Chang. © 1997 Karger Landes Systems.

••••••••••••••••••••••••••

Future Perspectives of Blood Substitutes

Where to Go from the Present First Generation Blood Substitutes?

Fluids for volume replacement have been in routine clinical use for many years. When fluid replacement alone is not enough, red blood cell or whole blood is the only next step right now. The present effort in the development of blood substitutes is to produce a substitute for red blood cells (Fig. 44).

At one time, it was thought that red blood cell substitutes had to be the same as blood in efficacy and safety. Are we correct in making this assumption? Most of us who have been involved in the research and development of red blood cell substitutes look at the first generation of blood substitutes as an intermediate step between volume replacement and red blood cells. In other words, the first generation blood substitute is not prepared to have the same efficacy and safety as red blood cells because it is something in–between. Furthermore, there are also different types of first generation blood substitutes such that each may be best used for a specific application. In this regard, we can learn much from the development of fluids used in volume replacement (Fig. 44).

The first type of fluid used for volume replacement did not start with the same efficacy and safety as plasma. For nearly a hundred years, we had saline that does not have even the same electrolyte composition as plasma. It is by far less effective than plasma. It was then improved to Ringer's lactate. Then it was further improved by adding colloid to Ringer's lactate for providing colloid osmotic pressure. However, it is very important to note that despite this type of increasing improvement, we did not discard saline or Ringer's lactate. Saline is

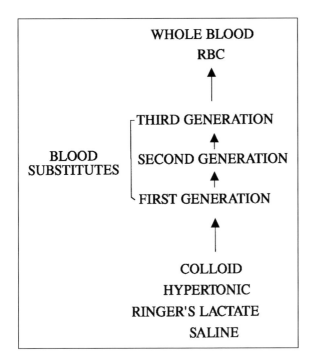

WHOLE BLOOD
RBC

THIRD GENERATION

BLOOD
SUBSTITUTES   SECOND GENERATION

FIRST GENERATION

COLLOID
HYPERTONIC
RINGER'S LACTATE
SALINE

Fig. 44. Progressions of different generations of fluid replacements and red blood cell substitutes. Reprinted with permission from Chang TMS. Artificial Cells, Blood Substitutes and Immobilization Biotechnology, an International Journal 1997; 25:1–24. Courtesy of Marcel Dekker Inc.

still being used for volume replacement in dialysis and in other situations. Furthermore, one does not always add colloids to Ringer's lactate. Ringer's lactate is used by itself most of the time and colloid is only needed in some special situations.

For blood substitutes, we now have several first generation blood substitutes that are already in phase I, II and III clinical trials with some nearly ready for routine clinical use in patients. This promising result should propel investigators to develop second and third generation products (Fig. 44). However, as for fluid replacement systems, it does not mean that once these are developed, we have to discard the first generation products. Unless the condition demands it, there is no need to use more expensive and more complicated second and third generation blood substitutes, if the first generation blood substitute can be used. Therefore, it is a matter of its use in specific situations.

Second Generation Blood Substitutes

General

Many improvements discussed earlier related to crosslinked hemoglobin or recombinant hemoglobin can be considered as second generation hemoglobin. This includes research on modifications of hemoglobin to change their affinity to nitric oxide, to change their oxygen affinity and other properties. The following are other important areas: S–nitrosothiols, reperfusion injury, preparations with antioxidant properties and crosslinked hemoglobin–SOD–catalase.

S–Nitrosothiols

The present notion is that the major role of hemoglobin in red blood cells and hemoglobin blood substitutes is to transport oxygen from the lung to the tissues. However, a recent exciting finding by Stamler's group [225] seems to suggest that hemoglobin in circulating red blood cells may also have an important role in the transport of nitric oxide (NO) and S–nitrosothiols (SNO). Very briefly, this starts with three simple facts: (1) Oxyhemoglobin has a higher affinity for SNO; (2) Deoxyhemoglobin has a higher affinity for NO; and (3) Lung is a good source of SNO.

When hemoglobin takes up oxygen in the lung it becomes oxyhemoglobin that has a high affinity for SNO (Fig. 45). It therefore also takes up SNO. This way, hemoglobin carries both oxygen and SNO in the arterial blood to the tissues. As oxyhemoglobin releases oxygen to the tissues, hemoglobin becomes deoxyhemoglobin. Since deoxyhemoglobin has a low affinity for SNO and a high affinity for NO, it releases SNO and picks up NO (Fig. 45). This moves to the lung. Here, deoxyhemoglobin picks up oxygen to become oxyhemoglobin and releases NO in exchange for SNO. Therefore, the cycle continues.

It has been proposed that this way, SNO has an important role in the control of vasoactivity [225]. If this theory is supported by other workers, it may have very important implications in the design of future generations of hemoglobin–based blood substitutes. Thus, besides the transport of oxygen, we may also have to look into the transport of NO and SNO by modified hemoglobin blood substitutes.

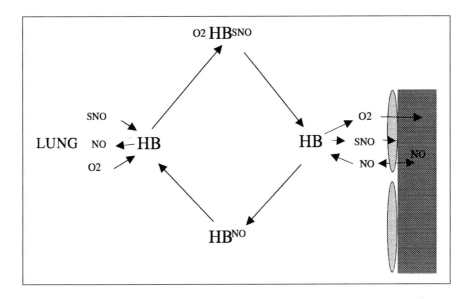

Fig. 45. This author's schematic representation of Stamler's theory of S–nitrosothiols and nitric oxide transport by hemoglobin in red blood cells. Reprinted with permission from Chang TMS. Artificial Cells, Blood Substitutes and Immobilization Biotechnology, an International Journal 1997; 25:1–24. Courtesy of Marcel Dekker Inc.

Reperfusion Injury

Red blood cells contain catalase, superoxide dismutase and other enzymes. However, modified hemoglobin blood substitutes are prepared using ultrapure hemoglobin without any enzyme systems. This is because of the need to remove all traces of endotoxin and other potential contaminants in the preparation of first generation blood substitutes. Many groups are analyzing the effects of modified hemoglobin on ischemic reperfusions and on nitric oxide. Some examples include those from the groups of Alayash [6, 7], Greenburg [245, 246], McKenzie [286], Messmer [341], Muldoon [308], Vercellotti [440, 442] and many others.

Superoxide dismutase and catalase play an important role in reperfusion injury. Lack of oxygen supply from hemorrhagic shock or other causes of inadequate circulation or oxygenation results in ischemia (Fig. 46). Ischemia stimulates the production of hypoxanthine. It also activates the enzyme xanthine oxidase. When the tissue is reperfused with oxygen, xanthine oxidase converts hypoxanthine into superoxide. By several mechanisms, superoxide results in the formation of oxygen radicals. Oxygen radicals can cause tissue injury (Fig. 46).

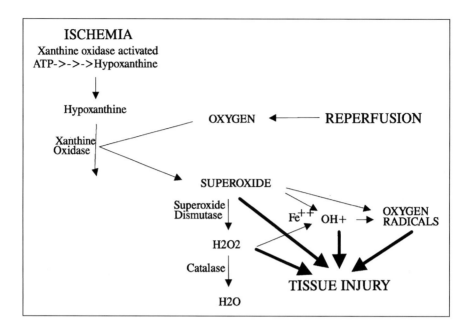

Fig. 46. Ischemic Reperfusion Injuries. Ischemia leads to accumulation of hypoxanthine and activation of xanthine oxidase. Reperfusion bringing oxygen resulted in superoxide formation. This and other resulting oxidants and oxygen radicals can cause tissue injury. Superoxide is removed to some extent by superoxide dismutase and catalase of red blood cells. First generation blood substitutes are ultrapure hemoglobin preparations not containing these enzymes. Reprinted with permission from Chang TMS. Artificial Cells, Blood Substitutes and Immobilization Biotechnology, an International Journal 1997; 25:1–24. Courtesy of Marcel Dekker Inc.

Enzymes in red blood cells help to prevent this to some extent. Thus, superoxide dismutase converts superoxide into hydrogen peroxide that is in turn converted by catalase into water and oxygen (Fig. 46). However, the present first generation modified ultrapure hemoglobin does not contain any of these enzymes. This may mean that there is potentially a higher chance for reperfusion injury when using blood substitutes prepared from ultrapure hemoglobin. In preparing second generation modified hemoglobin, we may want to go a step further to include enzymes or other antioxidants in the blood substitutes.

There are several approaches to counteract this potential problem. Biro's group [11, 25] suggested that activated polyglutamate polymerized hemoglobin may act as a scavenger of free iron. Hsia et al proposed the use of polynitroxylated hemoglobin with antioxidant activity [203]. Simoni et al prepared a "novel" hemoglobin based blood substitutes by modification of the hemoglobin molecule for the same reason [399–402]. Lemon et al constructed recombinant hemoglobin to alter the intrinsic rate of reactivity of hemoglobin for nitric oxide [259]. This is done by mutagenesis of the distal heme pockets. We have been studying the crosslinking of trace amounts of catalase and superoxide dismutase to hemoglobin [120–123, 353, 354, 358].

Crosslinked Hemoglobin–SOD–Catalase

In vitro study comparing polyHb–SOD–catalase to ultrapure polyhemoglobin shows the following results (Fig. 47) [120–123, 353, 354, 358]. PolyHb–SOD–catalase is much more effective in removing oxygen radicals and peroxides. It also stabilizes the crosslinked hemoglobin resulting in decreased oxidative iron and heme release. It also reduces the formation of methemoglobin during the preparation of polyhemoglobin.

Materials

The following materials were used: Xanthine oxidase (20 U/ml) and xanthine from ICN Biomedicals; Superoxide (bovine rbc, EC 1.15.1.1, 3000 units/ mg) and catalase (beef liver, EC 1.11.1.6, 65000 units/mg) from Boehringer Mannheim; Cytochrome c (horse heart, type III), 4–aminoantipyrine, horseradish peroxidase type IV (EC 1.11.1.7), ferrozine (0.85%), and the iron standard (500 g/dl) from Sigma; hemoglobin assay kit from StanBio Labs.

Preparation of PolyHb and PolyHb–SOD–Catalase

Polyhemoglobin (PolyHb) was prepared as described in the laboratory procedure earlier in this book. PolyHb–SOD–catalase was prepared as follows. Bovine hemoglobin (110 mg/ml), SOD (2 mg/ml), and catalase (20 mg/ml) were mixed in 0.1 M sodium phosphate buffer, pH 7.6 with the final ratio (as mg/ml) of Hb : SOD : catalase of 55 : 0.5 : 0.25. Following the addition of an initial amount of lysine–HCL (0.12 ml of 1.3 M/g Hb), gluteraldehyde (0.5 ml of 0.5 M/g Hb) was added to crosslink the protein mixture. The reaction was allowed to proceed for 1.5–2 hrs before being stopped by addition of excess lysine (0.78 ml of 2.0 M/g Hb). The resulting mixture was dialyzed against Ringer's lactate solution, then filtered through a 0.2 micron Nalgene filter. The hemoglobin concentration was measured. Molecular weight distribution analysis was

POLYHB-SUPEROXIDE DISMUTASE-CATALASE BUT NOT POLYHB

1. REMOVE OXYGEN RADICALS & PEROXIDES

2. DECREASE OXIDATIVE IRON & HEME RELEASE

3. DECREASE FORMATION OF METHB

Fig. 47. Comparison of Polyhemoglobin (PolyHb) with PolyHb–SOD–catalase. Reprinted with permission from Chang TMS. Artificial Cells, Blood Substitutes and Immobilization Biotechnology, an International Journal 1997; 25:1–24. Courtesy of Marcel Dekker Inc.

performed using gel filtration chromatography on a Sephadex G–200 column equilibrated with 0.1 M Tris–HCl, pH 7.5. The ratio of hemoglobin to SOD and catalase (as mg/ml) was 1 : 0.009 : 0.0045. The added SOD and catalase, therefore, did not significantly change the molecular weight distribution of the polyhemoglobin.

Monitoring Absorbance Spectra Following Oxidative Challenge

Hydrogen peroxide was added to PolyHb (10 mM) or PolyHb–SOD–catalase (10 mM), and the absorbance spectra (450–700 nm) were recorded over time. Following incubation with equimolar H_2O_2 (10 M), the spectral changes of PolyHb reflect the oxidation of ferrous (Fe^{2+})–heme producing ferric (Fe^{3+})–heme (Fig. 48). The absorbance spectra of PolyHb–SOD–catalase were minimally affected due to the elimination of H_2O_2. Similar results were recorded following oxidative challenge with exogenous O_2 via xanthine/xanthine oxidase.

Scavenging of Hydrogen Peroxide (H_2O_2)

Reaction volumes (3 ml) containing the horseradish peroxidase/4–amino-antipyrine/phenol reagent solution (1.2 ml), PolyHb or PolyHb–SOD–catalase (5 μM), water, and hydrogen peroxide were prepared. Identical mixtures containing additional water instead of H_2O_2 served as blanks. After allowing the mixture

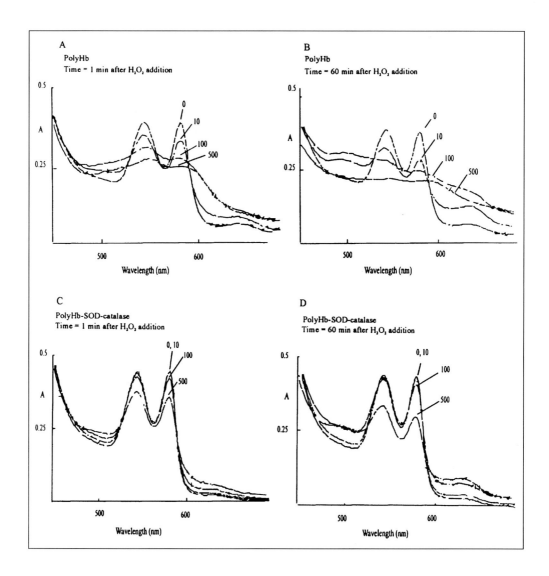

Fig. 48. Crosslinking hemoglobin with superoxide dismutase and catalase increases the stability of the polyhemoglobin in the presence of oxidizing agents. Reprinted with permission from D'Agnillo F, Chang TMS. Biomaterials, Artificial Cells and Immobilization Biotechnology 1993; 21:609–622. Courtesy of Marcel Dekker Inc.

to stand for 3 min. at 22°C, the absorbance at 505 nm was recorded. Hydrogen peroxide and the reagent solution participate in a peroxidase–catalyzed reaction to form a dye that can be measured at this wavelength. The results show that PolyHb–SOD–catalase was effective in scavenging hydrogen peroxide.

Scavenging of Superoxide (O_2-)

This is based on the reduction of cytochrome c by superoxide. Each reaction mixture (3 ml) contains xanthine (50 μM), cytochrome c (10 μM), and catalase (10 nM) in 50 mM potassium phosphate buffer containing 0.1 mM EDTA at pH 7.8. Free catalase was added to reaction mixtures to prevent interference resulting from the accumulation of H_2O_2. Each reaction mixture also contains either PolyHb (5 μM) or PolyHb–SOD–catalase (5 μM). Addition of 10 ml xanthine oxidase (4 U/ml) starts the reaction at 22°C. The rate of cytochrome c reduction was monitored at 550 nm with a Perkin–Elmer Lambda 4B Spectrophotometer. The molar extinction coefficient used for reduced cytochrome c was 2.1×10^4 M^{-1} cm^{-1}. The result is shown in Fig. 49. The initial rates of cytochrome c reduction were 0.56 ± 0.08 nmoles cyt. c/min for PolyHb–SOD–catalase compared to 2.13 ± 0.26 for PolyHb.

Iron Measurement

The first step was to incubate PolyHb (15 mM) or PolyHb–SOD–Catalase (15 mM) in hydrogen peroxide (total volume; 0.5 ml) for 60 min. at 37°C. Catalase is added to remove residual H_2O_2, then ascorbic acid (0.5 ml of 0.02%) was added and mixed for 5 min. Trichloroacetic acid (0.5 ml of 20%) was then added to precipitate proteins. The 1.5 ml mixture was centrifuged, and the supernatant (1 ml) was added to ammonium acetate buffer (0.45 ml) and ferrozine reagent (50 μl). The iron color complex was measured at 560 nm. The amount of iron released was calculated by measuring the absorbance of an iron standard (500 μg/dl) (0.5 ml), treated as described above, against blank (0.5 ml H_2O). ($A_{unknown}/A_{standard}$ * 500). With the addition of 500 mM of H_2O_2 37% of the total iron in PolyHb was "freed" (Fig. 50). For PolyHb–SOD–catalase less than 1% was released (Fig. 50).

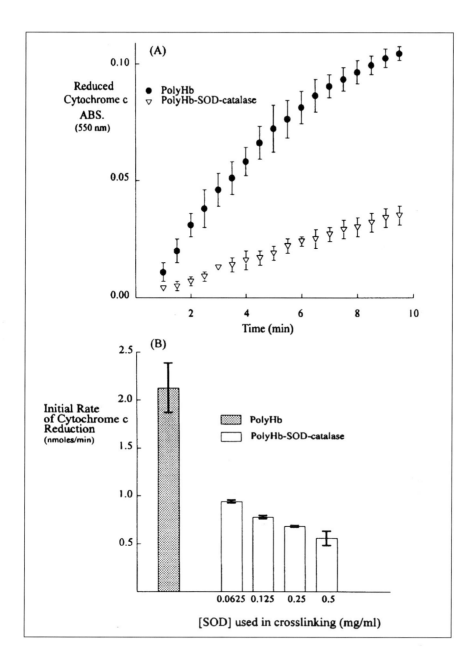

Fig. 49. PolyHb–SOD–catalase reduces superoxide in vitro. Superoxide level is followed by the level of reduced cytochrome c. Reprinted with permission from D'Agnillo F, Chang TMS. Biomaterials, Artificial Cells and Immobilization Biotechnology 1993; 21:609–622. Courtesy of Marcel Dekker Inc.

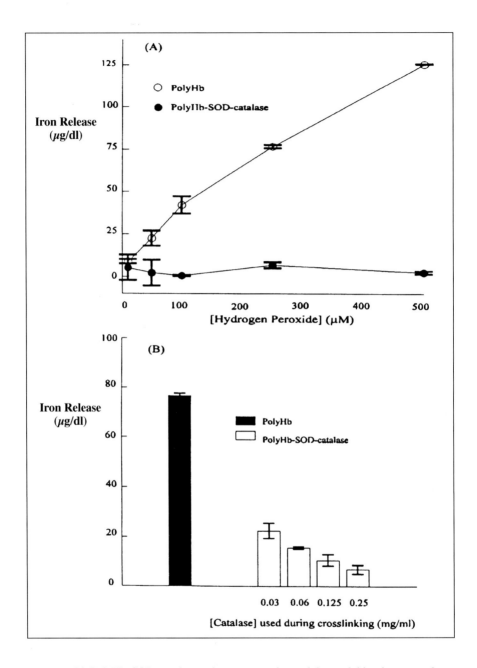

Fig. 50. PolyHb–SOD–catalase, when compared to polyhemoglobin, does not release significant amounts of iron in the presence of hydrogen peroxide. Reprinted with permission from D'Agnillo F, Chang TMS. Biomaterials, Artificial Cells and Immobilization Biotechnology 1993; 21:609–622. Courtesy of Marcel Dekker Inc.

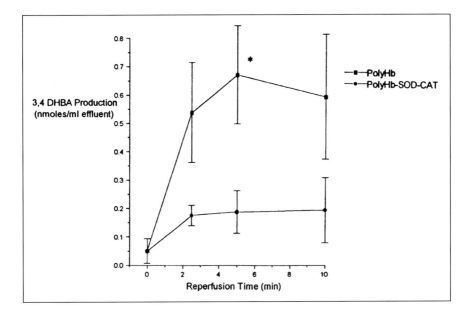

Fig. 51. Reperfusion of ischemia intestinal loop using Polyhemoglobin (PolyHb) or PolyHb–SOD–catalase. Effects on oxygen radicals as measured by effluent 3,4 dihydroxthenzoate. Reprinted with permission from Razack S, D'Agnillo F, Chang TMS. Artificial Cells, Blood Substitutes and Immobilization Biotechnology 1997; 25:181–192. Courtesy of Marcel Dekker Publisher Inc.

Reperfusion of Ischemic Intestine in Rats

We have carried out preliminary studies using ischemic reperfusion for both rat intestine [358] and rat hindlimbs [123]. The result of the intestinal ischemic reperfusion study is shown in Figure 51. Crosslinked ultrapure polyhemoglobin causes the formation of oxygen radicals as measured by an increase in 3,4 dihydroxybenzoate. This is significantly reduced when we used PolyHb–SOD–catalase for the reperfusion.

Third Generation Modified Hemoglobin Blood Substitute from Encapsulated Hemoglobin?

Crosslinked hemoglobins as described above are simpler and therefore are the first modified hemoglobins ready for clinical trials and eventually routine clinical use. However, crosslinked hemoglobin is only a partial substitute for red

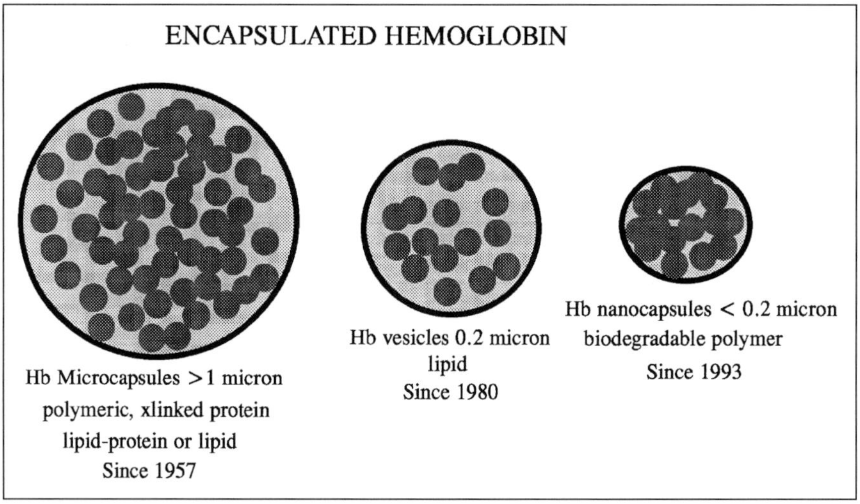

ENCAPSULATED HEMOGLOBIN

Hb Microcapsules >1 micron
polymeric, xlinked protein
lipid-protein or lipid
Since 1957

Hb vesicles 0.2 micron
lipid
Since 1980

Hb nanocapsules < 0.2 micron
biodegradable polymer
Since 1993

Fig. 52. Historical developments of encapsulated hemoglobin as artificial red blood cells.

blood cells. Since hemoglobin is not covered, it has to be ultrapure to avoid adverse reactions. Artificial red blood cells formed by encapsulation of hemoglobin and enzymes are more complete red blood cell substitutes. However, being more complete, they are also more complicated. Therefore, it takes longer to develop for clinical use. At present, artificial red blood cells are considered as second or third generation blood substitutes closer in properties to red blood cells.

There are three steps in the development of artificial red blood cells based on encapsulated hemoglobin (Fig. 52).

Original Microencapsulated Hemoglobin Artificial Red Blood Cells

The first study on microencapsulated hemoglobin or artificial red blood cells was reported by Chang in 1957 [47]. In this approach, synthetic membranes are used to replace the biological membranes of red blood cells. The resulting artificial red blood cells have an oxygen dissociation curve similar to red blood cells. The membrane used at this time was coated with a thin layer of butyl benzoate [47]. This retained 2,3–DPG inside. Here, a synthetic membrane replaces the natural red blood cell membrane. Further earlier studies by his group included the use of other synthetic polymers and crosslinked protein [48–52, 58–59]; membrane with surface charge [48–52, 58–59] and polysaccharide surface as

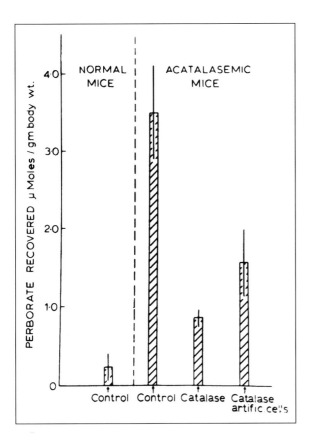

Fig. 53. Artificial cells containing hemoglobin and catalase for the removal of hydrogen peroxide in mice with congenital deficiency of catalase by Chang and Poznansky, 1968 [54]. Reprinted with permission from Chang TMS. Artificial Cells, Charles C. Thomas Publisher, 1972. Courtesy of copyright holder, TMS Chang.

sialic acid analogs [53, 56, 59]; lipid–protein and lipid–polymer [55, 59]. These artificial red blood cells do not have blood group antigens on the membrane. As a result, they do not form aggregates in the presence of blood group antibodies [59].

Hemoglobin inside the artificial red blood cells stays as tetramers. By proper adjustments of the membrane permeability, 2,3–DPG can be retained inside to improve oxygen release [47]. The resulting oxygen dissociation curve is similar to that of red blood cells (Fig. 5) [47]. Red blood cell enzymes, carbonic anhydrase [48] and catalase [54] in these microcapsules retained their activities.

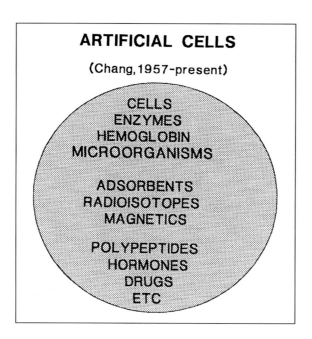

Fig. 54. Artificial cells containing different biologically active materials. Reprinted with permission from Chang TMS. Artificial Cells, Blood Substitutes and Immobilization Biotechnology 1994; 22: vii–xiv. Courtesy of Marcel Dekker Inc.

Encapsulated catalase can act as an antioxidant against the toxic effects of hydrogen peroxide. For example, we used acatalasemic mice with an inborn error of metabolism in their catalase enzyme. Figure 53 shows that the injection of hemoglobin artificial red blood cells containing catalase can lower the toxic peroxide levels in these animals [54]. Catalase from a heterologous source, unlike the free enzyme, did not induce an immunological reaction when given to immunized animals [348]. Artificial cells in the micron diameters range are being used extensively in other areas of biotechnology. These are administered by extracorporeal hemoperfusion or by implantation or orally for cell therapy, enzyme therapy, gene therapy, and drug delivery (Fig. 54) [78, 98].

The single major problem for using encapsulated hemoglobin in the micron range for blood substitutes is rapid removal after intravenous infusion. We found [49, 59] that they have to be less than 1 micron in diameter to avoid being trapped in the lung capillaries. Even then, they are removed by the reticuloendothelial system. We attempted to look at why red blood cells circulate for so long by using neuraminidase to remove sialic acid from the red blood cell membrane.

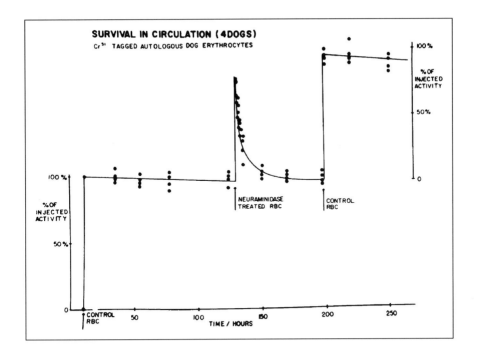

Fig. 55. In 1965, Chang reported that removal of sialic acid from red blood cell membrane by treating with neuraminidase resulted in their rapid removal. Three intravenous infusions: (1) Infusion of radioisotope labeled control autologous red blood cells and followed for 120 minutes; (2) Infusion of red blood cells with sialic acid removed by neuraminidase; and (3) Infusion of control autologous red blood cells. From the original 1965 figure by Chang [49]. Reprinted with permission from Chang TMS. Ph.D. Thesis, McGill University, 1965. (Also reprinted in Artificial Cells, Charles C. Thomas Publisher, 1972). With permission of the author.

We reported this result in 1965 showing that removal of sialic acid resulted in the rapid removal of RBC from the circulation (Fig. 55) [49, 50, 59]. This observation led us to prepare artificial red blood cells with modifications of surface properties. This included the use of other synthetic polymers and crosslinked protein (Fig. 6) [48–52, 58–59]; membrane with surface charge (Fig. 56) [48–52, 58–59] and polysaccharide surface as sialic acid analogs [53, 56, 59]; lipid–protein and lipid–polymer [55, 59]. Some of these improved the circulation time. However, the circulation time was still not enough for practical applications.

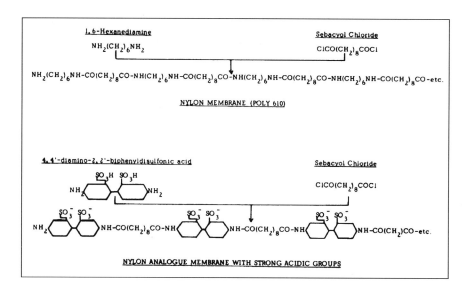

Fig. 56. Examples of negatively charged membranes to change the surface properties. Reprinted from the original 1965 figure by Chang [49] with permission from Chang TMS. Ph.D. Thesis, McGill University, 1965. (Also reprinted in Artificial Cells, Charles C. Thomas Publisher, 1972).

Submicron Encapsulated Hemoglobin Inside Lipid Vesicles

This is important area will be covered in details by Rudolph in a later chapter for the second volume of this book.

Djordjevich and Miller reported in 1980 that they could prepare 0.2 micron diameters lipid membrane vesicles encapsulating hemoglobin [59]. This substantially increased the circulation time, although the circulation time was still rather short. Many groups have since carried out research to improve the preparation and the circulation time. These groups include those of Beissinger et al [18]; Domokos, Schmidt [133]; Farmer [146]; Gaber [160]; Hunt and Burnette [221, 212]; Miller et al [292]; Mobed and Chang [296–301]; Nishiya [323–325]; Phillips et al [342, 343]; Rudolph [101, 111, 372–374]; Szeboni et al [407–410]; Takahashi [412–414]; Tsuchida et al [428–431, 415, 416]; Usuuba [435–437]. Modifications of surface properties including surface charge and the use of sialic acid analogs have further improved the circulation time. The average half–time in the circulation is now more than 30 hours. It is possible to replace most of the red blood cells in rats with these artificial red blood cells.

Some examples of more recent studies are as follows. Szebeni et al [410] are studying the interaction of hemoglobin lipid vesicles with human complement using an in–vitro screening method similar to the one developed earlier by Chang [80]. Takaori and Fukui [413] used hemoglobin lipid vesicles for massive hemorrhage in animal studies. Usuba's group [435–437] used hemoglobin lipid vesicles for total cardiopulmonary bypass and studied the effects on hemodynamics and blood gas transport in canine hemorrhagic shock. Other studies included the incorporation of enzymatic reduction systems of methemoglobin by Ogata et al [328]. Tsuchida's group is attempting to solve the problem of methemoglobin formation by using artificial reduction systems as recently reported by Takeoka et al [416].

The groups of Rudolph in the U.S.A. and Tsuchida in Japan have made extensive progress. They are collaborating with many groups using their preparations. Studies by several groups show that there are no adverse changes in the histology of brain, heart, kidneys and lungs of experimental animals. Effects on the reticuloendothelial system have been studied by a number of groups. Recent examples include those of Beach et al [17], Dittman et al [131], Phillips et al [242] and others. Preclinical studies are being conducted.

More details are available regarding the comparison of hemoglobin lipid vesicles and hemoglobin nanocapsules in the next section. Dr. Rudolph is writing a detailed chapter in the second volume of this book to cover this large and important area.

Biodegradable Polymeric Hemoglobin Nanocapsules

Just as success in crosslinked hemoglobin stimulates research into next generation crosslinked hemoglobin, this is also the case in encapsulated hemoglobin. This is perhaps the time for the next step toward a further generation of encapsulated hemoglobin. For instance, one can look into how to further improve the following:
1. Increase stability in storage and after infusion.
2. Decrease the potential effects of lipid on the reticuloendothelial systems.
3. Avoid lipid peroxidation.
4. Solve the problem of methemoglobin formation.

We are using our background in biodegradable polymer encapsulation started here by this author in 1976 [61]. Polylactides and polyglycolides are degraded in the body into water and carbon dioxide. The rate of degradation can be adjusted by changes in molecular weight and type of polymer or copolymer. It can also vary with particle size. We are now using this to prepare biodegradable polymer membrane hemoglobin [87, 102, 459–462] to have a mean diameter of

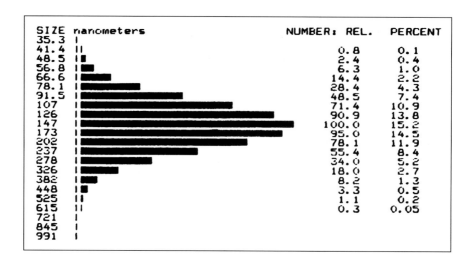

Fig. 57. Size distribution of biodegradable polymeric hemoglobin nanocapsules. Reprinted with permission from Yu WP, Chang TMS. Artificial Cells, Blood Substitutes and Immobilization Biotechnology, an International Journal 1996; 24:169–184. Courtesy of Marcel Dekker Inc.

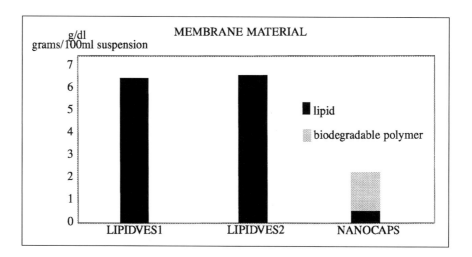

Fig. 58. Amount of membrane materials in each 100 ml suspension. Hemoglobin lipid vesicles (LIPIDVES) compared to that of biodegradable polymeric hemoglobin nanocapsules (NANOCAPS). Reprinted with permission from Chang TMS. Artificial Cells, Blood Substitutes and Immobilization Biotechnology, an International Journal 1997; 25:1–24. Courtesy of Marcel Dekker Inc.

FATE OF POLYLACTIC ACID IN BODY

polylactide ---> lactic acid ---> $CO_2 + H_2O$

BIODEGRADABLE HB NANOCAPSULES: 500 ml

Total amount of lactic acid:	83 mEq/week = 12 mEq/day
Resting body lactic acid production:	1000-1400 mEq/day
Ability of body to breakdown lactic acid:	7680 mEq/day

Fig. 59. Fates of polylactic acid membrane of hemoglobin nanocapsules in the body. Reprinted with permission from Chang TMS. Artificial Cells, Blood Substitutes and Immobilization Biotechnology, an International Journal 1997; 25:1–24. Courtesy of Marcel Dekker Inc.

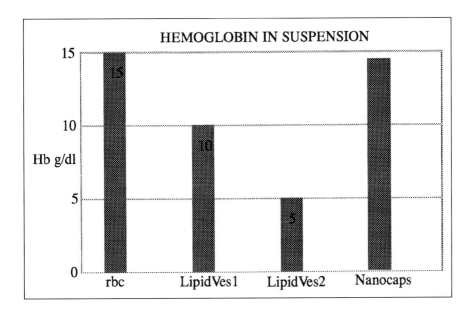

Fig. 60. Hemoglobin in each 100 ml suspension of whole blood (rbc), hemoglobin lipid vesicles (LipidVes) and hemoglobin nanocapsules (Nanocaps). Reprinted with permission from Chang TMS. Artificial Cells, Blood Substitutes and Immobilization Biotechnology, an International Journal 1997; 25:1–24. Courtesy of Marcel Dekker Inc.

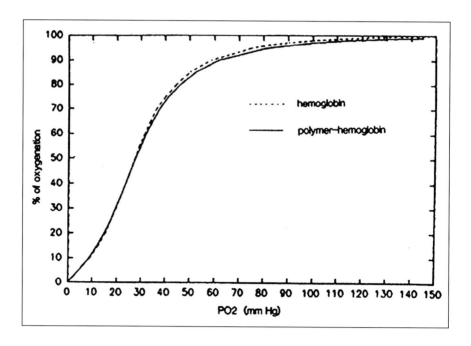

Fig. 61. Oxygen dissociation curve of bovine hemoglobin in the free form and inside hemoglobin nanocapsules. No significant differences in P_{50}. There is also no significant differences in cooperativity. Reprinted with permission from Yu WP, Chang TMS. Artificial Cells, Blood Substitutes and Immobilization Biotechnology, an International Journal 1996; 24:169–184. Courtesy of Marcel Dekker Publisher Inc.

between 80–200 nanometers (Fig. 57). The membrane material is made up mostly of biodegradable polymer. Since polymer is stronger and is also porous, less membrane material is required (Fig. 58). Polylactide is degraded in the body into lactic acid, water and carbon dioxide. For a 500 ml suspension, the total lactic acid produced is 83 mEq (Fig. 59). This is far less than the normal resting–body lactic acid production (1000–1400 mEq/day). This is equivalent to 1% of the capacity of the body to breakdown lactic acid (7080 mEq/day).

The content of hemoglobin can match that of red blood cells (Fig. 60). Bovine hemoglobin after encapsulation has the same P_{50}, Bohr and Hill coefficients (Fig. 61). Superoxide dismutase and catalase can also be included with the hemoglobin [85]. We have used our background in artificial cells containing multienzyme cofactor recycling systems [63, 65, 68, 77, 98, 183–185, 463–465] to help solve the problem of methemoglobin formation in encapsulated hemoglobin lipid vesicles. Applying this to hemoglobin nanocapsules may help to solve

PERMEABILITY		
	GLUCOSE	NADH
RBC	YES	NO
LIPIDVES1	NO	NO
LIPIDVES2	NO	NO
NANOCAPS	YES	ADJUSTABLE

Fig. 62. Comparison of membrane permeability of red blood cells (RBC), hemoglobin lipid vesicles (LIPIDVES) and hemoglobin nanocapsules (NANOCAPS). Permeability to glucose is needed for multienzyme reactions. Reprinted with permission from Chang TMS. Artificial Cells, Blood Substitutes and Immobilization Biotechnology, an International Journal 1997; 25:1–24. Courtesy of Marcel Dekker Inc.

the problem related to methemoglobin reductase. Since lipid vesicles are not permeable to glucose, the required glucose is added in high concentrations into the lipid vesicles [414, 416]. In nanocapsules, the biodegradable polymeric membranes can be made permeable to glucose and other molecules (Fig. 62) [102, 464]. This allows us to prepare hemoglobin nanocapsules containing the methemoglobin reductase system to function (Fig. 63) [102, 464]. External glucose can diffuse into the nanocapsules. Products of the reaction can diffuse out and therefore do not accumulate in the nanocapsules to inhibit the reaction. Preliminary in vitro study shows that this can convert methemoglobin to hemoglobin.

Animals have been infused with one-third the total blood volume. Studies are being carried out to increase the circulating time of these hemoglobin nanocapsules.

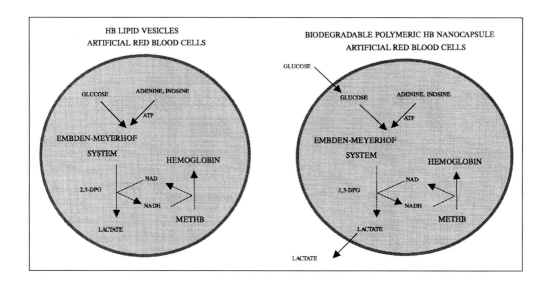

Fig. 63. Hemoglobin nanocapsule containing red blood cell hemolysate allows the red blood cell enzymes and cofactor to regenerate cofactors and convert methemoglobin to hemoglobin. Left: hemoglobin lipid vesicles. Right: Hemoglobin nanocapsules that can use external glucose from blood. Superoxide dismutase and catalase have also been enclosed into the nanocapsules. This has resulted in a more complete red blood cell. Reprinted with permission from Chang TMS. Artificial Cells, Blood Substitutes and Immobilization Biotechnology, an International Journal 1997; 25:1–24. Courtesy of Marcel Dekker Inc.

Chapter 9

Blood Substitutes: Principles, Methods, Products and Clinical Trials,
by Thomas Ming Swi Chang. © 1997 Karger Landes Systems.

....

Should There Be a Priority on Blood Substitutes for National Blood Supply Policies?

In any national policy on blood supply, there is an urgent need to include blood substitutes as a priority area for research and development towards clinical application. Let us look at what this means.

What Happens When There Is No National Priority?

As early as 1957, encapsulated hemoglobin to form artificial red blood cells was carried out [47]. In 1964, crosslinking hemoglobin to form polyhemoglobin was carried out [48]. In 1968, intramolecularly crosslinking hemoglobin was shown to prevent hemoglobin from entering the kidney to cause toxic effects [38]. Thus, the basic ideas of modified hemoglobin, although very crude, were all there in the 1960s. However, with no national priority and no public interest at that time, research and developments in other areas of artificial cells was emphasized [78, 93, 98].

In the early 1980s, as discuss earlier, stoma–free hemoglobin and perfluorochemicals were shown not to work well. It is know retrospectively that at that time, the defense departments of several countries started to carry out research on modified hemoglobin. However, there was little academic, industrial or public interest.

Concentrated effort to develop blood substitutes for public use was only seriously started after 1986 because of public concerns regarding HIV from donor blood. Unfortunately, a product cannot be ready for clinical use without years of research and development followed by clinical trials. More than 10 years have passed since 1986 and it will take another two years for one product and many more years for others to become available for routine use in patients. Why should this happen when the basic ideas of encapsulated hemoglobin and crosslinked hemoglobin were already there in the 1960s? Had there been national policies 30 years ago to develop blood substitutes, we would have had blood substitutes ready in 1986 during the HIV crisis in donor blood. As it is, the public has continued to be exposed to the rare though potential hazard of HIV in donor blood since 1986. Even in 1997, blood substitutes will still not be ready for routine clinical use for one to two years. Even then, these are first generation blood substitutes that are only partial substitutes for donor blood. There is still much to be done.

What Happens When There Is a National Priority?

An US national committee in early 1990 arrived at the conclusion that research and development in modified hemoglobin should be considered a priority area. The NIH implemented this very shortly after. Three major 5 years program grants have been awarded to three universities. The NIH has also sent out RFA for 8 to 10 smaller 5 year grants. In addition, there were also major navy and army research efforts. This also stimulated the many ongoing industrial efforts on modified hemoglobin. Now the following US companies are working on modified hemoglobin: Apex Bioscience; Baxter; Biopure; Enzon; Northfield and Somatogen. Two other companies are in perfluorochemicals: Alliance and Hemogen. Two of these companies are already into phase III clinical trials. The US effort on all fronts will put them in the forefront of research, development and supply of modified hemoglobin.

What Happens When the National Priority Is Short–Term?

Unfortunately, with the present economic situation around the world, the national priority discussed above was short lived. When the industries had developed the first generation blood substitutes for clinical trials, support for non-industrial and academic research started to decrease markedly in the U.S. and even more so in other countries. Perhaps it is thought that the industry can support all the research in this area. Unfortunately, most of the industries in this area are small, venture capital-supported companies. They have very limited capitals and have problems developing and carrying out the very expensive clinical trials for their first generation blood substitutes. Meanwhile, the second and third generation blood substitutes are left to random support here and there. We are now moving on the same path as before and repeating our mistake. One should not leave any doubts that even if we start with maximal efforts immediately, it will take many more years before the second and third generation blood substitutes would be ready just for initial testing in humans.

Why should we go further than the present first generation blood substitute products now in clinical trial? One can imagine similar questions being asked in many other areas. One can also imagine what would have happened if we had stopped at penicillin and did not develop further generations of antibiotics. What would have happened if we had stopped at the method of detecting hepatitis B for donor blood and did not go further to develop methods to detect hepatitis C and HIV?

What Needs to Be Implemented?

To avoid delay in the routine clinical use of second and third generation blood substitutes, we need urgently to include blood substitutes in the national priority. With the present economic problems in most countries, granting agencies tend to divert their resources to national priorities. What is left is then divided among the numerous other nonpriority areas, including blood substitutes. Herein lies the problem for research and development on blood substitutes since the 1960s. We therefore need to urgently establish blood substitutes as one of the national priorities. Only this way can this area receive in urgently needed priority support. In addition, we need urgently to establish some mechanisms, e.g., national and international network. This will bring together the investigators, the industries, the Red Cross Societies, the blood agencies, the research agencies, federal and provincial agencies and private groups that are most affected by decisions in this area. It is only by doing this with our limited resources that we can effectively carry out the following activities:

1. Conduct research, development, industrial production and clinical trials on blood substitutes.

2. The universities have to train the next generation of well–qualified investigators who can be responsible for research and development. What is also equally important is urgently to supply experienced scientists and physician–scientists who can decide, evaluate, judge and use blood substitutes clinically.

3. Basic research in the university contributes new novel ideas for the next generation of blood substitutes for development by the industry. First generation blood substitutes will shortly be available for routine clinical application. However, this is just the first step and much more research is needed for the second and third generation blood substitutes.

··

References

1   Agisaka K, Iwashita Y. Modification of human hemoglobin with polyethylene glycol: A new candidate for blood substitute. Biochem Biophys Res Commun 1980; 97:1076.

2   Agishi T, Funakoshi Y, Honda H, Yamagata K, Kobayashi M, Takahashi M. (Pyridoxalated hemoglobin) – (Polyoxyethylene) Conjugate solution as blood substitute for normothermic whole body rinse–out. Biomaterials, Artificial Cells and Artificial Organs 1988; 16:261–270.

3   Agishi T, Sonda K, Nakajima I, Fuchinoue S, Honda H, Uga HS, Liu H, Teraoka S, Ota K. Modified hemoglobin solution as possible perfusate relevant to organ transplantation. Biomaterials, Artificial Cells and Immobilization Biotechnology 1992; 20:539–544.

4   Agishi T, Funakoshi Y, Fujita S, Fuchinoue S, Ota K. Abstract volume for VI International Symposium on Blood Substitutes. Artificial Cells, Blood Substitutes & Immobilization Biotechnology, an International Journal 1996; Vol 24, issue 4.

5   Alayash AI, Fratantoni JC. Effects of hypotermic conditions on the oxygen carrying capacity of crosslinked hemoglobins. J Biomaterials, Artificial Cells and Immobilization Biotechnology 1992; 20:259–262.

6   Alayash AI, Brockner Ryan BA, Fratantoni JC. Hemoglobin–based oxygen carriers: structural alterations that affect free radical formation. Biomaterials, Artificial Cells and Immobilization Biotechnology 1992; 20:277–282.

7   Alayash A. Abstract volume for VI International Symposium on Blood Substitutes. Artificial Cells, Blood Substitutes & Immobilization Biotechnology, an International Journal 1996; Vol 24, issue 4.

8   Amberson WR. Blood substitute. Biol Rev 1937; 12:48.

9   Amend J, Ou C, Ryan–MacFarlane C, Anderson PJ, Amend N, Biro JP. Systemic responses to SFHS–infusion in hemorrhaged dogs. Artificial Cells, Blood Substitutes & Immobilization Biotechnology, an International Journal 1996; 24:19–34.

10 Anderson PJ, Ning J, Biro GP. Clearance of differentially labelled infused hemoglobin and polymerized hemoglobin from dog plasma and accumulation in urine and selected tissues. Biomaterials, Artificial Cells and Immobilization Biotechnology 1992; 20:781–788.

11 Anderson PJ, Vaccani JP, Biro GP. Abstract volume for VI International Symposium on Blood Substitutes. Artificial Cells, Blood Substitutes & Immobilization Biotechnology, An International Journal 1996; Vol 24, issue 4.

12 Bakker JC, Bleaker WK, Van der Plas J. Hemoglobin interdimerically crosslinked with NEPLP. Biomaterials, Artificial Cells and Artificial Organs 1988; 16:635–636.

13 Bakker JC, Berbers WAM, Bleeker WK, den PJ Boer, Biessels PTM. Characteristics of crosslinked and polymerized hemoglobin solutions. Biomaterials, Artificial Cells and Immobilization Biotechnology 1992; 20:233–242.

14 Bakker JC, Bleaker WK. Blood substitutes based on modified hemoglobin. Vox Sang 1994; 67(suppl):139–142.

15 Baron BJ, Scalea TM. Acute blood loss. Emerg Med Clin North Am 1996; 14:35–55.

16 Barve A, Sen AP, Gulati A. Dose Response effect of diaspirin crosslinked hemoglobin (DCLHb) on systemic hemodynamics and regional blood circulation in rats. Artificial cells, Blood Substitutes & Immobilization Biotechnology, an International Journal 1997; 25:75–84.

17 Beach MC, Morley J, Spiryda L, Weinstock SB. Effects of liposome encapsulated hemoglobin on the reticuloendothelial system. Biomaterials, Artificial Cells and Artificial Organs 1988; 16:635–636.

18 Beissinger RL, Farmer MC, Gossage JL. Trans Am Soc Artif Inter Organs 1986; 32:58.

19 Bekyarova G, Yankova T, Galunska B. Increased antioxidant capacity, suppression of free radical damage and erythrocyte aggrerability after combined application of alpha–tocopherol and FC–43 perfluorocarbon emulsion in early postburn period in rats. Artificial Cells, Blood Substitutes & Immobilization Biotechnology, an International Journal 1996; 24:629–641.

20 Benesch R, Benesch RE, Yung S, Edalji R. Hemoglobin covalently bridged across the polyphosphate binding site. Biochem Biophys Res Commun 1975; 63:1123.

21 Biessels PTM, Berbers GAM, Broeders GCJM, Landsvater R, Huisman J, Bleeker WK, Bakker JC. Detection of membrane fragments in hemoglobin solutions. Biomaterials, Artificial Cells and Immobilization Biotechnology 1992; 20:439–442.

22 Biro GP. Fluorocarbons in the resuscitation of hemorrhage. Perflurochemical Oxygen Transport 1985; 23(1):143.

23 Biro GP. Perfluorocarbon–based red blood cell substitutes. Transfus Med Rev 1993; 7:84–95.

24 Biro GP. Central hemogdynamics and blood flow distribution during infusion of perflubron emulsion or its vehicle. Artificial Cells, Blood Substitutes and Immobilization Biotechnology, an International Journal 1994; 22:1343–1353.

25 Biro GP, Ou C, Ryan–MacFarlane C, Anderson PJ. Oxyradical generation after resuscitation of hemorrhagic shock with blood or stromafree hemoglobin. Artificial cells, Blood substitutes & Immobilization Biotechnology, an International Journal 1995; 23:631–645.

26 Bleeker WK, Zappeij LM, den Boer PJ, Agterberg JA, Rigter GMM, Bakker JC. Evaluation of the immunogenicity of polymerized hemoglobin solutions in a rabbit model. Artificial cells, Blood substitutes & Immobilization Biotechnology, an International Journal 1995; 23:461–468.

27 Bleeker W, Agterberg J, La Hey E, Rigter G, Zappeij L, Bakker J. Hemorrhagic disorders after administration of gluteraldehyde–polymerized hemoglobin. In: Winslow RM, Vandegriff KD, Intaglietta M, eds. Blood Sunstitutes: New Challenges. Boston: Birkhäuser, 1996:112–123.

28 Omitted in proof.

29 Bolin RB, Geyer RP, Nemo GJ, eds. Advances in Blood Substitute Research. New York: Alan R Liss Inc. 1983.

30 Bonhard K. Hemoglobin preparations for perfusion and infusion problems of large–scale production. Biomaterials, Artificial Cells and Artificial Organs 1988; 16:85–92.

31 Bowersox JC, Hess JR. Trauma and military applications of blood substitutes. Artificial Cells, Blood Substitutes and Immobilization Biotechnology, an International Journal 1994; 22:145–159.

32 Bowes MP, Burhop KE, Zivin JA, Abraham DJ. Diaspirin crosslinked hemoglobin improves neurological outcome following reversible but not irreversible CNS ischemia in rabbits. Stroke 1994; 25:2253–2257.

33 Braun RD, Linsenmeier RA, Goldstick TK. New perfluorocarbon emulsion improves tissue oxygenation in retina. J Appl Physiol 1992; 72:1960–1968.

34 Bucci E, Fronticelli B. Hemoglobin tetramers stabilized with polyaspirins. Biomaterials, Artificial Cells and Immobilization Biotechnology 1992; 20:243–252.

35 Bucci E, Fronticelli C, Razynska A, Militello V, Koehler R, Urbaitis. Hemoglobin tetramers stabilized with polyaspirins. In: Chang TMS, ed. Blood Substitutes & Oxygen Carrier. New York: Marcel Dekker Publisher, 1992:76–85.

36 Bucci E, Kwansa H, Razynska A, O'Heame. Abstract volume for VI International Symposium on Blood Substitutes. Artificial Cells, Blood Substitutes and Immobilization Biotechnology, an International Journal 1996; Vol 24, issue 4.

37 Bucci E, Razynska A, Kwansa H, Matheson–Urbaitis B, O'Hearne M, Ulatowski JA, Koehler RC. Production and characteristics of an infusible oxygen–carrying fluid based on hemoglobin intramolecualrly crosslinked with sebacic acid. J Lab Clin Med 1996; 128:146–153.

38 Bunn HF, Jandl JH. The renal handling of hemoglobin. Trans Assoc Am Physicians 1968; 81:147.

39 Burhop KE, Farrell L, Nigro C, Tan D, Estep T. Effects of intravenous infusions of diaspririn crosslinked hemoglobin (DCLHb) on sheep. Biomaterials, Artificial Cells and Immobilization Biotechnology 1992; 20:581–586.

40 Burkard ME, Van Liew IID. Oxygen transport to tissue by persistent bubbles:theory and simulations. J Appl Physiol 1994; 77:2874–2978.

41 Campbell J, Chang TMS. Enzymatic recycling of coenzymes by a multi–enzyme system immobilized within semipermeable collodion microcapsules. Biochim Biophys Acta 1975; 397:101–109.

42 Campbell J, Chang TMS. The recycling of NAD^+ (free and immobilized) within semipermeable aqueous microcapsules containing a multi–enzyme system. Biochem Biophys Res Commun 1976; 69(2):562–569.

43 Campbell J, Chang TMS. Immobilized multienzyme systems and coenzyme requirements: Perspectives in biomedical applications. In: Chang TMS, ed. Biomedical Applications of Immobilized Enzymes and Proteins. Vol 2. New York: Plenum Press, 1977:281–302.

44 Campbell J, Chang TMS. Microencapsulated multi–enzyme systems as vehicles for the cyclic regeneration of free and immobilized coenzymes. Enzymes Engineering 1978; 3:371–377.

45 Caspirin, speaker, VI International Symposium on Blood Substitutes. 1996.

46 Cerny LC, Barnes B, Fisher L, Anibarro M, Ho N. A starch–hemoglobin resuscitative compound. Artificial Cells, Blood Substitutes and Immobilization Biotechnology, an International Journal 1996; 24:153–160.

47 Chang TMS. Hemoglobin corpuscles. Report of a research project of B.Sc. Honours Physiology, McGill University 1–25, 1957. Medical Library, McIntyre Building, McGill University, 1957 (reprinted in "30th Anniversary in Artificial Red Blood Cell Research" Biomaterials, Artificial Cells and Artificial Organs 1988; 16:1–9.

48 Chang TMS. Semipermeable microcapsules. Science 1964; 146(3643):524.

49 Chang TMS. Ph.D. Thesis, McGill University.1965.

50 Chang TMS, MacIntosh FC. Semipermeable aqueous microcapsules. Proceedings of the XIII International Congress of Physiological Sciences, Tokyo, Japan 1965.

51 Chang TMS. Semipermeable aqueous microcapsules ("Artificial cells"): with emphasis on experiments in an extracorporeal shunt system. Trans Am Soc Artif Int Organs 1966; 12:13–19.

52 Chang TMS, MacIntosh FC, Mason SG. Semipermeable aqueous microcapsules: I. Preparation and properties. Can J Physiol Pharmacol 1966; 44:115–128.

53 Chang TMS, Johnson LJ, Ransome OJ. Semipermeable aqueous microcapsules: IV. Nonthrombogenic microcapsules with heparin–complexed membranes. Can J Physiol Pharmacol 1967; 45:705–715.

54 Chang TMS, Poznansky MJ. Semipermeable microcapsules containing catalase for enzyme replacement in acatalsaemic mice. Nature 1968; 218(5138):242–45.

55 Chang TMS. Lipid–coated spherical ultrathin membranes of polyer of crosslinked protein as possible cell membrane models. Fed Proc Am Soc Exp Biol 1969; 28:461(abstract).

56 Chang TMS. "Nonthrombogenic Microcapsules": U.S. Patent 3,522,346, 1970.

57 Chang TMS. Stabilization of enzyme by microencapsulation with a concentrated protein solution or by crosslinking with glutaraldehyde. Biochem Biophys Res Com 1971; 44:1531–1533.

58 Chang TMS, MacIntosh FC, Mason SG. Encapsulated hydrophilic compositions and Methods of Making them. Canadian Patent 873,815, 1971.

59 Chang TMS. Artificial cells. Monograph. Charles C Thomas, Springfield, IL, 1972.

60 Chang TMS. Microcapsule artificial kidney: including updated preparative procedures and properties. Kidney Int 1976; 10:S218–S224.

References

117

61 Chang TMS. Biodegradable semipermeable microcapsules containing enzymes, hormones, vaccines, and other biologicals. J Bioengineering 1976; 1:25–32.

62 Chang TMS. Microencapsulation of enzymes and biologicals. Methods in Enzymology 1976; XLIV:201–217.

63 Chang TMS, ed. Biomedical Applications of Immobilized Enzymes & Proteins. Volumes I and II. New York: Plenum Publishing Corporation, 1977.

64 Chang TMS. Artificial red blood cells. Trans Am Soc Artif Int Organs 1980; 26:354–357.

65 Chang TMS, Yu YT, Grunwald J. Artificial cells immobilized multienzyme systems and cofactors. Enzyme Engineering 1982; 6:451–561.

66 Chang TMS. Artificial cells in medicine and biotechnology. Applied Biochemistry and Biotechnology 1984; 10:5–24.

67 Chang TMS, ed. Microencapsulation and Artificial Cells. Clifton, New Jersey, USA: Humana Press, 1984.

68 Chang TMS. Artificial cells with cofactor regenerating multienzyme systems. Methods in Enzymology 1985; 112:195–203.

69 Chang TMS. Modified Hemoglobin as Red Blood Cell Substitutes. Biomaterials, Artificial Cells and Artificial Organs 1987; 15:323–328.

70 Chang TMS. The use of modified hemoglobin as an oxygen carrying blood substitute. Transfusion Medicine Review 1987; 3:213–218.

71 Chang TM, ed. Blood Substitutes and Oxygen Carriers. New York: Marcel Dekker Publisher, 1992:784.

72 Chang TMS, Varma R. Pyridoxalated heterogenous and homologous polyhemoglobin and hemoglobin: systemic effects of replacement transfusion in rats previously received immunising doses. Biomaterials, Artificial Cells and Artificial Organs 1987; 15:443–452.

73 Chang TMS. Red blood cells substitutes: Microencapsulated hemoglobin and crosslinked hemoglobin including pyridoxalated polyhemoglobin and conjugated hemoglobin. Biomaterials, Artificial Cells and Artificial Organs 1988; 16:11–29.

74 Chang TMS, Geyer R,. eds. Blood Substitutes. New York: Marcel Dekker Publisher, 1988.

75 Chang TMS, Varma R. Immunological and systemic effects of transfusions in rats using pyridoxalated hemoglobin and polyhemoglobin from homologous and heterogenous sources. Biomaterials, Artificial Cells and Artificial Organs 1988; 16:205–215.

76 Chang TMS. The use of modified hemoglobin as an oxygen carrying blood substitute. Transfusion Medicine Review 1989; 3:213–218.

77 Chang TMS, Daka J. Removal of bilirubin by the pseudoperoxidase activity of immobilized hemoglobin. U.S. Patent No. 4820416, 1989.

78 Chang TMS. Artificial cells. In: Dulbecco R, ed. Encyclopedia of Human Biology. Vol 1. San Diego, CA: Academic Press Inc., 1990:377–383.

79 Chang TMS. Modified hemoglobin: Endotoxin and safety studies. Biomaterials, Artificial Cells and Artificial Organs 1990; 18:vii–viii.

80 Chang TMS, Lister C. A screening test of modified hemoglobin blood substitute before clinical use in patients – based on complement activation of human plasma. Biomaterials, Artificial Cells and Artificial Organs 1990; 18(5):693–702.

81 Chang TMS, Varma R. Effects of Ringer lactate, albumin, stroma–free hemoglobin, o–raffinose polyhemoglobin, and whole blood on lethal hemorrhagic shock in rats. Biomaterials, Artificial Cells and Immobilization Biotechnology 1991; 19:368–571.

82 Chang TMS. Blood substitutes based on modified hemoglobin prepared by encapsulation or crosslinking. Biomaterials, Artificial Cells and Immobilization Biotechnology 1992; 20:154–174.

83 Chang TMS, Lister C. An in vitro screening test for modified hemoglobin to bridge the gap between animal safety studies and clinical use in patients. Biomaterials, Artificial Cells and Immobilization Biotechnology 1992; 20:(in this issue).

84 Chang TMS, Varma R. Effect of a single replacement of Ringer lactate, 7% albumin, hypertonic saline/dextran, stroma–free hemoglobin, o–raffinose polyhemoglobin, or whole blood, on the long term survival of fully conscious rats with lethal hemorrhagic shock after 67% acute blood loss. Biomaterials, Artificial Cells and Immobilization Biotechnology 1992; 20:503–510.

85 Chang TMS, Yu WP. Biodegradable polymer membrane containing hemoglobin as potential blood substitutes. British Provisional Patent No. 92194265, September 14, 1992.

86 Chang TMS, Lister C, Nishiya T, Varma R. Effects of different methods of administration and effects of modifications by microencapsulation, crosslinkage or PEG conjugation on the immunological effects of homologous and heterologous hemoglobin. Biomaterials, Artificial Cells and Immobilization Biotechnology 1992; 20:611–618.

87 Chang TMS, Yu WP. Biodegradable polymer membrane containing hemoglobin as potential blood substitutes. British Provisional Patent No. 92194265, September 14, 1992.

88 Chang TMS. Safety studies of modified hemoglobin as oxygen carrying blood substitute. Hematologic Pathology 1993; 7:47–52.

89 Chang TMS, Lister CW. Screening test for modified hemoglobin blood substitute before use in human. U.S. Patent No. 5,200,323, April 6, 1993.

90 Chang TMS, Lister CW. Use of finger–prick human blood samples as a more convenient way for in vitro screening of modified hemoglobin blood substitutes for complement activation: a preliminary report. Biomaterials, Artificial Cells and Immobilization Biotechnology 1993; 21:685–690.

91 Chang TMS, Lister C, Wong LT, Er SS. The use of a preclinical screeningtest based on C3a activation of human plasma in industrial production of o–raffinose polymerized hemoglobin. #H93 Abstract, Vth International Symposium on Blood Substitutes, San Diego, March 17–20, 1993.

92 Chang TMS. Efficacy of rbc substitutes. Artificial Cells, Blood Substitutes and Immobilization Biotechnology 1994; 22:ii–iii.

93 Chang TMS. Artificial Cell including blood substitutes and biomicroencapsulation: from ideas to applications. Artificial Cells, Blood Substitutes and Immbolization Biotechnology 1994; 22:vii–xiv.

94 Chang TMS, (editor–in–chief). Abstracts on Blood Substitutes from the XI Congress of the International Society for Artificial Cells, Blood Substitutes & Immobilization Biotechnology. Artificial Cells, Blood Substitutes and Immobilization Biotechnology, an International Journal 1994; 22(5): A75–A177.

95 Chang TMS, Lister C. Assessment of Blood Substitutes: II. In vitro complement activation of human plasma and blood for safety studies in research, development, industrial production and preclinical analysis. Artificial Cells, Blood Substitutes & Immobilization Biotechnology, an International Journal 1994; 22:159–170.

96 Chang TMS, Reiss JG, Winslow R (guest editors). Symposium volume on "Blood Substitutes: General." Artificial Cells, Blood Substitutes and Immobilization Biotechnology, An International Journal 1994; 22:123–360.

97 Chang TMS, Varma R. Assessment of Blood Substitutes: I. Efficacy studies in anesthetized and conscious rats with loss of 1/3, 1/2 and 2/3 blood volume. Artficial Cells, Blood Substitutes and Immbolization Biotechnology, an International Journal 1994; 22:159–170.

98 Chang TMS. Artificial cells with emphasis in bioencapsulation for biotechnology and medicine. Biotechnology Annual Review 1995; 1:267–296.

99 Chang TMS. Crosslinked hemoglobin being well into clinical trials, increasing research efforts are now on a second generation red blood cell substitute based on encapsulated hemoglobin. Artificial Cells, Blood Substitutes and Immobilization Biotechnology, an International Journal 1995; 23:257–262.

100 Chang TMS, Weinstock S, eds. Blood Substitutes. Special issue, Artificial Cells, Blood Substitutes and Immbilization Biotechnology, an International Journal 1995; 23:257–459.

101 Chang TMS. Past, present and future perspectives on the 40th anniversary of hemoglobin–based red blood cell substitutes. Artificial Cells, Blood Substitutes and Immobilization Biotechnology, an International Journal 1996; 24:ix–xxiv.

102 Chang TMS, Yu WP. Biodegradable polymer membrane containing hemoglobin for blood substitutes. U.S.A. Patent. 1996.

103 Chang TMS. Recent and Future Developments of Modified Hemoglobin and Microencapsulated Hemoglobin as Red Blood Cell Substitutes. Artificial Cells, Blood Substitutes and Immobilization Biotechnology, an International Journal 1997; 25:1–24.

104 Chang TMS, Yu WP. Biodegradable polymer membrane containing hemoglobin for blood substitutes. U.S.A. Patent 1997.

105 Chang TMS, Greenberg AG, Tsuchida E, eds. Blood Substitutes. Special issue, Artificial Cells, Blood Substitutes and Immbolization Biotechnology, an International Jounral 1997; 25:1–242.

106 Chapman KW, Snell SM, Jesse RG, Morano JK, Everse J, Winslow RM. Pilot scale production of pyrogen–free modified human hemoglobin for research. Biomaterials, Artificial Cells and Immobilization Biotechnology 1992; 20:415–422.

107 Chapman KW, Keipert PC, Graham HA. Abstract volume for VI International Symposium on Blood Substitutes. Artificial Cells, Blood Substitutes & Immobilization Biotechnology, an International Journal, 1996; Vol 24, issue 4.

108 Chi OZ, Lu X, Wei HM, Williams JA, Weiss HR. Hydroxyethyl starch solution attenuates blood–brain disruption caused by intracarotid injection of hyperosmolar mannitol in rats. Anesth Analg 1996; 83:336–341.

109 Clark LC Jr, Gollan F. Survival of mammals breathing organic liquids equilibrated with oxygen at atmospheric pressure. Science 1966; 152:1755.

110 Cliff RO, Ligler F, Goins B, Hoffmann PM, Spielberg H, Rudolph AS. Liposome encapsulated hemoglobin. Biomaterials, Artificial Cells and Immobilization Biotechnology 1992; 20:.

111 Cliff RO, Kwasiborski V, Rudolph AS. a comparative study of the accurate measurement of endotoxin in liposome encapsulated hemoglobin 1995; 23:331–336.

112 Cole DJ, Przybelski RJ, Schell RM, Martin RD. Diaspirin crosslinked hemoglobin (DCLHb™) does not affect the anesthetic potency of isoflurane in rats. Artificial Cells, Blood Substitutes & Immobilization Biotechnology, An International Journal 1995; 23:89–99.

113 Cole DJ, Drummond JC, Patel PM, Nary JC, Applegate RL. Effect of oncotic pressure of diaspirin crosslinked hemoglobin (DCLHb) on brain injury after temporary focal cerebral ischemia in rats. Anesth Analg 1996; 83:342–347.

114 Cole DJ, McKay L, Jacobsen WK, Drummond JC, Patel PM. Effect of subarachnoid administration of a–a diaspirin crosslinked hemoglobin on cerebral blood flow in rats. Artificial Cells, Blood Substitutes & Immobilization Biotechnology, An International Journal 1997; 25:95–104.

115 Cole DJ, Nary JC, Drummond JC, Patel PM, Jacobsen WK. a–a diaspirin crosslinked hemoglobin, nitric oxide, and cerebral ischemic injury in rats. Artificial Cells, Blood Substitutes & Immobilization Biotechnology, An International Journal 1997; 25:141–152.

116 Conover C, Lejeune L, Linberg R, Shum K, Shorr RGL. Transitional vacuole formation following a bolus infusion of PEG–hemoglobin in the rat. Artificial Cells, Blood Substitutes & Immobilization Biotechnology, An International Journal 1996; 24:599–611.

117 Conover CD, Malatesta P, Lejeune L, Chang CL, Shorr RG. The effects of hemodilution with polyethylene glycol bovine hemoglobin (PEG–Hb) in a conscious porcine model. J Investig Med 1996; 44:238–246.

118 Daka JN, Chang TMS. Bilirubin removal by the pseudoperoxidase activity of free and immobilized hemoglobin and hemoglobin co–immobilized with glucose oxidase. Biomaterials Artificials Cells and Artificial Organs 1989; 17:553–562.

119 Daka JN, Sipehia R, Chang TMS. Enhanced oxidation of bilirubin by an immobilized tri–enzyme system of glucose oxidase, bilirubin oxidase and horseradish peroxidase. Biochim Biophys Acta 1989; 991:487–489.

120 D'Agnillo F, Chang TMS. Crosslinked hemoglobin–superoxide dismutase–catalase scavenges oxygen–derived free radicals and prevents methemoglobin formation and iron release. Biomaterials, Artificial Cells and Immobilization Biotechnology 1993; 21:609–622.

121 D'Agnillo F, Chang TMS. Abstract volume for VI International Symposium on Blood Substitutes. Artificial Cells, Blood Substitutes and Immobilization Biotechnology, an International Journal 1996; Vol 24, issue 4.

122 D'Agnillo F, Chang TMS, Modified hemoglobin blood substitute from Crosslinked hemoglobin–superoxide dismutase–catalase. US patent 1997.

123 D'Agnillo F, Chang TMS. Reduction of hydroxyl radical generation in a rat hindlimb model of ishemia–reperfusion injury using crosslinked hemoglobin–superoxide dismutase–catalase. Artifi-

cial Cells, Blood Substitutes and Immobilization Biotechnology, an International Journal 1997; 25:163–180.

124 Davies A, Magnum A. Media release. Hemosol Inc., January 1995.

125 Davey MG. In Blood Services in Canada: The Canadian Red Cross Society in book on Blood Substitutes. Chang TMS, Geyer RP, eds. New York: Marcel Dekker Publisher, 1989:51–53.

126 Dellacherie E et al. Hemoglobin linked to polymeric effectors as red blood cell substitutes. Biomaterials, Artificial Cells and Immobilization Biotechnology 1992; 20:309–318.

127 Detsch O, Heesen M, Muhling J, Thiel A, Backmann–Mennenga B, Hempelmann G. Isovolemic hemodilution with hydroxyethylstarch has no effect on somatosensory evoled potentials in healthy volunteers. Acta Anesthesiol Scand 1996; 40:665–670.

128 DeVenuto F, Zuck TF, Zegna AI, Moores WY. Characteristics of stroma–free hemoglobin prepared by crystallization. J Lab Clin Med 1977; 89:509.

129 DeVenuto F, Zegna AI. Blood exchange with pyridoxalated–polymerized hemoglobin. Surg Gynecol Obstet 1982; 155:342.

130 Dietz NM, Joyner MJ, Warner MA. Blood substitutes: fluids, drugs, or miracle solutions? Anesth Analg 1996; 82:390–405.

131 Dittmer J, Prusty S, Ichikura T, Pivacek L, Giorgio A, Valeri CR. Intravascular retention and distribution of DBBF crosslinked stroma–free hemoglobin in the mouse. Biomaterials, Artificial Cells and Immobilization Biotechnology 1992; 20:751–756.

132 Djordjevich L, Miller IF. Synthetic erythrocytes from lipid encapsulated hemoglobin. Exp Hematol 1980; 8:584.

133 Domokos G, Jopski B, Schmidt KH. Prepration, properties and biological function of liposome encapsulated hemoglobin. Biomaterials, Artificial Cells and Immobilization Biotechnology 1992; 20:345–354.

134 Dudziak R, Bonhard K. The development of hemoglobin preparations for various indications. Anesthesist 1980; 29:181.

135 Dunlap E, Farrell L, Nigro C, Estep T, Marchand G, Burhop K. Resuscitation with diaspirin crosslinked hemoglobin in a pig model of hemorrhagic shock. Artificial Cells, Blood Substitutes and Immobilization Biotechnology, an International Journal 1995; 23:39–61.

136 Dupuis NP, Kusumoto T, Robinson MF, Liu F, Menon K, Teicher BA. Restoration of tumor oxygenation after cytotoxic therapy by a perflubron emulsion/carbogen breathing. Artificial Cells, Blood Substitutes and Immobilization Biotechnology, an International Journal 1995; 23:423–429.

137 Edwards CM, Lowe KC, Trabelsi H, Lucas P, Cambon A. Novel Fluorinated Surfactants for Perfluorochemical Emulsification: Biocompatibility Assessments of Glycosidic and Polyol Derivatives. Artificial Cells, Blood Substitutes and Immobilization Biotechnology, an International Journal 1997; (in press).

138 Edwards CM, Lowe KC, Rohlke W, Geister U, Reuter P, Meinert H. Effects of a Novel Perfluorocarbon Emulsion on Neutrophil Chemiluminescence in Human Whole Blood In Vitro. Artificial Cells, Blood Substitutes and Immobilization Biotechnology, an International Journal 1997; (in press).

139 Eisman JM, Schnaare RL. The Formation of a Crosslinked Carboxyhemoglobin Membrane at an Organic–Aqueous Interface. Artificial Cells, Blood Substitutes and Immobilization Biotechnology, an International Journal 1996; 24:185–196.

140 Eldrige J, Russell R, Christenson R et al. Liver function and morphology after resuscitation from severe hemorrhagic shock with hemoglobin solutions or autologous blood. Crit Care Med 1996; 24:663–671.

141 Estep TN, Gonder J, Bornstein I, Young S, Johnson RC. Immunogenicity of diaspirin crosslinked hemoglobin solutions. Biomaterials, Artificial Cells and Immobilization Biotechnology 1992; 20:603–610.

142 Faassen A, Sundby SR, Panter SS, Condie RM, Hedlund BE. Hemoglobin: lifesaver and an oxidant. How to tip the balance. Biomaterials, Artificial Cells and Artificial Organs 1988; 16:93–104.

143 Faithfull NS. Oxygen delivery from fluorocarbon emulsions – aspects of convective and diffusive transport. Biomaterials, Artificial Cells and Artificial Organs 1992; 20:797–804.

144 Faithfull NS. Mechanisms and efficacy of fluorochemical oxygen transport and delivery. Artificial Cells, Blood Substitutes and Immobilization Biotechnology (Winslow R, guest editor) 1994; 22:687–694.

145 Faivre B, Labaeye V, Menu P, Labrude P, Vigneron C. Assessment of dextran 10–benzene tetracarboxlate–hemoglobin, an oxygen carrier, using guinea pig isolated bowel model. Artificial Cells, Blood Substitutes and Immobilization Biotechnology, an International Journal 1995; 23:495–504.

146 Farmer. MC, Rudolph AS, Vandegriff KD, Havre MD, Bayne SA, Johnson SA. Lipsome–encapsulated hemoglobin: oxygen binding properties and respiratory function. Biomaterials, Artificial Cells and Artificial Organs 1988; 16:289–299.

147 Farmer MC, Ebeling A, Marshall T, Hauck W, Sun CS, White E, Long Z. Validation of virus inactivation by heat treatment in the manufacture of diaspirin crosslinked hemoglobin. J Biomaterials, Artificial Cells and Immobilization Biotechnology 1992; 20:429–434.

148 Fattor TJ, Mathews AJ. Abstract volume for VI International Symposium on Blood Substitutes. Artificial Cells, Blood Substitutes and Immobilization Biotechnology, an International Journal 1996; Vol 24, issue 4.

149 Feola M, Gonzalez H, Canizaro PC, Bingham D. Development of bovine stroma–free hemoglobin solution as a blood substitute. Surgery Gynecology and Obstetrics 1983; 157:399.

150 Feola M, Simoni J, Ruc Tran DVM, Canizaro PC. Mechanisms of toxicity of hemoglobin solutions. Biomaterials, Artificial Cells and Artificial Organs 1988; 16:217–226.

151 Feola M, Simoni J. Biocompatibility of hemoglobin solutions: The effects of contaminants, hemoglobin and hemoglobin derivatives. Biomaterials, Artificial Cells and Immobilization Biotechnology 1991; 19(2):382.

152 Flaim SF, Hazard DR, Hogan J, Peters RM. Characterization and mechanism of side–effects of Imagent BP (highly concentrated fluorocarbon emulsion) in swine. Invest Radiol 1991; 26(Suppl):S125–S128(discussion).

153 Frankel HL, Nguyen HB, Shea–Donohue T, Aiton LA, Ratigan J, Malcolm DS. Diaspirin crosslinked hemoglobin is efficacious in gut resuscitation as measured by a GI tract optode. J Trauma 1996; 40:231–240; discussion 241.

154 Fratantoni JC. Points to consider in the safety evaluation of hemoglobin based oxygen carriers. Transfusion 1991; 31:369–371.

155 Fratantoni JC. Points to consider on efficacy evaluation of hemoglobin and perfluorocarbon based oxygen carriers. Transfusion 1994; 34:712–713.

156 Fratantoni JC. Demonstration of efiicay of a therapeutic agent. In: Winslow Rm, Vandergriff KD, Intaglietta M, eds. Blood Substitutes: Physiologival Basis of Efficacy. Boston: Birkhäuser, 1996.

157 Frey L, Messmer K. Transfusion therapy. Current Opin Anaesthesiol 1996; 9:183–187.

158 Freyburger G, Dubreuil M, Boisseau MR, Janvier G. Rheological properties of commonly used plasma substitutes during preoperative normovolaemic acute haemodilution. Br J Anaesth 1996; 76:519–525.

159 Fronticelli C, Brinigar W, Grycznski Z, Bucci E. Engineering of low oxygen affinity hemoglobins. Biomaterials, Artificial Cells and Immobilization Biotechnology 1991; 19:385–.

160 Gaber BP, Farmer MC. Encapsulation of hemoglobin in phospholipid vesicles: preparation and properties of a red blood cell surrogate. Prog Clin Biol Res 1984; 165:179–190.

161 Garrioch et al. Crit Care Med 1996; 24: A39.

162 Garcia–Sepulcre ME, Carnicer F, Mauri M, Prieto A, Perez–mateo M. Increased plasma endothelin in liver cirrhosis and response to plasma volume expansion [letter]. Am J Gastroenterol 1996; 91:2452–2453.

163 Gennaro M, Mohan C, Ascer E. Perfluorocarbon emulsion prevents eicoasanoid release in skeletal muscle ischemia and reperfusion. Cardiovasc Surg 1996; 4:399–404.

164 Geyer RP, Monroe RG, Taylor K. Survival of rats totally perfused with a fluorocarbon–detergent preparation. In: Norman JC, Folkman J, Hardison WG, Rudolf LE, Veith FJ, eds. Organ Perfusion and Preservation. New York: Appleton Century Crofts, 1968:85–96.

165 Geyer RP, Bolin, Nemo GJ (eds), Advances in Blood Substitute Research. New York: Alan R. Liss Inc., [year]:1–468.

166 Geyer RP. Blood–replacement preparations. In: Kirk–Othmer, ed. Encyclopedia of Chemical Technology. Supplement volume, 3rd ed. John Wiley & Sons, Inc., 1984.

167 Golubev AM, Magomedov MA, Leonteva TA, Guseinov TS. The biomicroscopic evaluation of the blood microcirculatory bed in the mesentary of the small intestine during experimental perftoran infusion under normovolemia. Morfologiia 1996; 109:36–40.

168 Goodin T. Results of a Phase I clinical trial of a 40 v/v% emulsion of HM351 (Oxyfluor™) in healthy volunteers. 2nd annual IBC conference on blood substitutes and related products. Washington DC, 1994.

169 Goodin TH, Grossbard EB, Kaufman RJ, Richard TJ, Kolata RJ, Allen JS, Layton TE. A Perfluorochemical emulsion for prehospital resuscitation of experimental hemorrhagic shock: a prospective, randomized controlled study. Crit Care Med 1994; 22:680–689.

170 Gotoh K, Morioka T, Nishi K. Effects of pyridoxalated hemoglobin polyoxyethylene conjugate (PHP) on pulmonary vascular responsiveness to various vasoactive substances in isolated perfused rat lungs. Biomaterials, Artificial Cells and Immobilization Biotechnology 1992; 20:721–722.

171 Gould SA, Sehgal LR, Sehgal HL, Moss GS (Northfield Co). Clinical Experience with human polymerized hemoglobin #H13, Abstract, Vth International Symposium on Blood Substitutes, San Diego, March 17–20, 1993.

172 Gould SA, Sehgal LR, Sehgal HL, Toyooka E, Moss GS. Clinical experience with human polymerized hemoglobin (abstract). Transfusion 1993; 33:(suppl 9S)60S.

173 Gould SA. Comments on Phase II clinical trial using polyhemoglobin at the XI Congress, International Society for Artificial Cells, Blood Substitutes & Immobilization Biotechnology, Boston, U.S.A. 1994.

174 Gould SA, Moore EE, Moore FA, Haenel J et al. The clinical utility of human polymerized hemoglobin as a blood substitute following trauma and emergent surgery. J Trauma 1995; 39:157.

175 Gould SA, Sehgal LR, Sehgal HL, Moss GS. The development of hemoglobin solutions as red cell substitutes: Hemoglobin solutions. Transfu Sci 1995; 16:5–17.

176 Gould SA, Moss GS. Clinical development of human polymerized hemoglobin as a blood substitute. World J Surg 1996; 20:1200–1207.

177 Greenburg AG, Hayashi R, Krupenas I. Intravascular persistence and oxygen delivery of pyridoxalated stroma–free hemoglobin during gradations of hypotension. Surgery 1979; 86:13.

178 Greenburg AG, Kim HW. Evaluating new red cell substitutes: A critical analysis of toxicity models. Biomaterials, Artificial Cells and Immobilization Biotechnology 1992; 20:575–581.

179 Greenburg AG. Nitrosyl hemoglobin formation in–vivo after intravenous administration of a hemoglobin–based oxygen carrier in endotoxemic rats. Artificial Cells, Blood Substitutes and Immobilization Biotechnology, an International Journal 1995; 23:271–276.

180 Greenburg AG. Use of donor blood in the surgical setting: potentials for application of blood substitutes. Artificial Cells, Blood Substitutes and Immobilization Biotechnology, an International Journal 1997; 25:25–30.

181 Greenwald RB, Pendri A, Martinez A, Gilbert C, Bradley P. PEG Thiazolidine–2–thoone, a novel reagent for facile protein modification: Conjugation of bovine hemoglobin. Bioconj Chem 1996; 7:638:641.

182 Griffiths E, Cortes A, Gilbert N et al. Hemoglobin based blood substitutes and sepsis. Lancet 1995; 345:158–160.

183 Grunwald J, Chang TMS. Immobilization of alcohol dehydrogenase, malic dehydrogenase and dextran–NAD+ within nylon–polyethyleneimine microcapsules: preparation and cofactor recycling. J Molecular Catalysis 1981; 11:83–90.

184 Grunwald J, Chang TMS. Nylon polyethyleneimine microcapsules for immobilizing multienzymes with soluble dextran–NAD+ for the continuous recycling of the microencapsulated dextran–NAD+. Biochem Biophys Res Commun 1978; 81(2):565–570.

185 Grunwald J, Chang TMS. Continuous recycling of NAD+ using an immobilized system of collodion microcapsules containing dextran–NAD+, alcohol dehydrogenase, and malic dehydrogenase. J Applied Biochem 1979; 1:104–114.

186 Gudi SRP, Clark CB, Frangos JA. Fluid flow rapidly activates G proteins in human endothelial cells. Involvement of G proteins in mechanicochemical signal transduction. Circ Res 1996; 79:834–839.

187 Gulati A, Sharma AC, Burhop KE. Effect of stroma–free hemoglobin and diasprin crosslinked hemoglobin on the regional circulation and systemic hemodynamics. Life Sci 1994; 55:827–837.

188 Gulati A, Sharma AC, Singh G. Role of endothelin in the cardiovascular effects of disapirin crosslinked and stroma reduced hemoglobin. Crit Care Med 1996; 24:137–147.

189 Hanzawa K, Ohzeki H, Moro H, Eguchi S, Nakajima T, Makifuchi T, Miyashita K, Nishiura M, Naritomi H. Effects of partial blood replacement with pyridoxalated hemoglobin polyoxyethylene conjugate solution on transient cerebral ischemia in gerbil. Artificial Cells, Blood Substitutes and Immobilization Biotechnology, an International Journal 1997; 25:105–114.

190 Hertzman CM, Keipert PE, Chang TMS. Serum antibody titers in rats receiving repeated small subcutaneous injections of hemoglobin or poly–hemoglobin: a preliminary report. Int J Artif Organs 1986; 9:179–182.

191 Hess JR, MacDonald VW, Winslow RM. Dehydration and Shock: An animal model of hemorrhagic and resuscitation of battlefield injury. Biomaterials, Artificial Cells and Immobilization Biotechnology 1992; 20:499–502.

192 Hess JR, Macdonald VW, Brinkley WW. Systemic and pulmonary hypertension after resuscitation with cell free hemoglobin. J Appl Physiol 1993; 74:1769–1778.

193 Hess JR. Review of Modified Hemoglobin Research at Letterman: Attempts to Delineate the Toxicity of Cell–Free Tetrameric Hemoglobin. Artificial Cells, Blood Substitutes and Immobilization Biotechnology, an International Journal 1995; 23:277–289.

194 Hobbhahn J, Vogel H, Kothe N, Brendel W, Keipert P, Jesch F. Hemodynamics and transport after partial and total blood exchange with pyridoxalated polyhemoglobin in dogs. Acta Anesth Scand 1985; 29:537–543.

195 Hodakowski GT, Page RD, Harringer W, Jacobs EE Jr, LaRaia PJ, Svizzero T, Guerrero JL, Austen WG, Vlahakes GJ. Ultra–pure polyemerized bovine hemoglobin blood substitute: effects on the coronary circulation. Biomaterials, Artificial Cells and Immobilization Biotechnology 1992; 20:669–672.

196 Hoffman SJ, Looker DL, Roehrich JM et al. Expression of fully functional tetrameric human hemoglobin in escherichia coli. Proc Natl Acad Sci USA 1990; 87:8521–8525.

197 Hogan MC, Willford DC, Keipert PE, Faithfull NS, Wagner PD. Increased plasma O_2 solubility improves O_2 uptake of in situ dog muscle working maximally. J Appl Physiol 1992; 74:2470–2475.

198 Holland PV. Viral infections and the blood supply. N Engl J Med 1996; 334:1734:1735.

199 Horiuchi T, Ohta Y, Hashimoto K, Dohi T, Uechi M. Machine perfusion of isolated kidney at 37°C using pyridoxalated hemoglobin–polyoxyethylene (PHP) solution, UW solution and its combination. Biomaterials, Artificial Cells and Immobilization Biotechnology 1992; 20:549–557.

200 Hsai JC. Pasteurizable, freeze–driable hemoglobin–based blood substitute. U.S. Patent 857,636, August 15, 1989.

201 Hsia JC. o–Raffinose polymerized hemoglobin as red blood cell substitute. Biomaterial Artificial Cells and Immobilization Biotechnology 1991; 19:402.

202 Hsia JC, Song DL, Er SS, Wong LTL. Pharmacokinetic studies on a raffinose–polymerized human hemoglobin in the rat. Biomaterials, Artificial Cells and Immobilization Biotechnology 1992; 20:587–596.

203 Hsia JC. Abstract volume for VI International Symposium on Blood Substitutes. Artificial Cells, Blood Substitutes and Immobilization Biotechnology, an International Journal 1996; Vol. 24, issue 4.

203 Hsia N, Everse J. The cytotoxic activities of human hemoglobin and diaspirin crosslinked hemoglobin. Artificial Cells, Blood Substitutes and Immobilization Biotechnology, an International Journal 1996; 24:533–551.

205 Hogan MC, Kurdak SS, Richardson RS, Wagner PD. Partial substitution of red blood cells with free hemoglobin solution does not improve maximal O_2 uptake of working in situ dog muscle. Adv Exp Med Biol 1994; 361:375–378.

206 Hughes GS, Jacobs E (Upjohn USA) The hemodynamic response to hemopure, a polymerized bovine hemoglobin solution, in normal subjects #H15 Abstract, Vth International Symposium on Blood Substitutes, San Diego, March 17–20, 1993.

207 Hughes GS, Francom SF, Antal EJ et al. Hematologic effects of a novel hemoglobin–based oxygen carrier in normal male and female subjects. J Lab Clin 1995; 126:444–451.

208 Hughes GS, Jr, Yancey EP, Albrecht R et al. Hemoglobin–based oxygen carrier preserves submaximal exercise capacity in humans. Clin Pharmacol Ther 1995; 58:434–443.

209 Hughes GS, Francom SF, Antal EJ, Adams WJ, Locker PK, Yancey EP, Jacobs EE. Effects of a novel hemoglobin–based oxygen carrier on percent oxygen saturation as determined with arterial blood gas analysis and pulse oximetry. Ann Emerg Med 1996; 27:164–169.

210 Hughes GS, Jr, Antal EJ, Locker PK, Francom SF, Adams WJ, Jacobs EE. Physiology and pharmacokinetics of a novel hemoglobin–based oxygen carrier in humans. Crit Care Med 1996; 24:756–764.

211 Hunt AC, Burnette RR. Neohemocytes. In: Bolin RB, Geyer RP, Nemo GJ, eds. Advances in Blood Substitute Research. New York: Alan R Liss Inc, 1983:59–70.

212 Hunt AC, Burnette RR, MacGregor RD et al. Science 1985; 230:1165.

213 Ilan E, Morton P, Chang TMS. Bovine hemoglobin anaerobically reacted with divinyl sulfone: A potential source for hypothermic oxygen carriers. Biomaterials, Artificial Cells and Immobilization Biotechnology 1992; 20:246–255.

214 Ilan E, Morton P, Chang TMS. Bovine hemoglobin anaerobically reacted with divinyl sulfone: A potential source for hypothermic oxygen carriers. Biomaterials, Artificial Cells and Immobilization Biotechnology 1992; 20:263–276.

215 Ilan E, Morton P, Chang TMS. The anaerobic reaction of bovine hemoglobin with divinyl sulfone: Structural changes and functional consequences. Biochimica and Biophysica Acta 1993; 1163:257–265.

216 Ilan E, Chang TMS. Human hemoglobin anaerobically reacted with divinyl sulfone: A source for oxygen carriers. Artificial Cells, Blood Substitutes and Immobilization Biotechnology (Winslow R, guest editor) 1994; 22:687–694.

217 Intaglietta M. Hemodilution and blood substitutes. Artificial Cells, Blood Substitutes and Immobilization Biotechnology (Winslow R, guest editor) 1994; 22:137–144.

218 Intaglietta M, Johnson PC, Winslow RM. Microvascular and tissue oxygen distribution. Cardiovasc Res 1996; 32:632–643.

219 Ivanistski GR, Vorobev SI. Organization of mobile structures in blood stream:basis for the function of perfluorocarbon artificial blood. Biofizika 1996; 41:178–190.

220 Iwashita Y, Yabuki A, Yamaji K, Iwasaki K, Okami T, Hirati C, Kosaka K. A new resuscitation fluid "stabilized hemoglobin" preparation and characteristics. Biomaterials, Artificial Cells and Artificial Organs 1988; 16:271–280.

221 Iwashita Y. Relationship between chemical properties and biological properties of pyridoxalated hemoglobin–polyoxyethylene. Biomaterials, Artificial Cells and Immobilization Biotechnology 1992; 20:299–308.

222 Jacobs E Jr, Hughes GS. Update on clinical aspects of polymerized bovine hemoglobin (HBOC–201). Artficial Cells, Blood Substitutes & Immbolization Biotechnology, An International Journal 1996; 24(abstracts issue):357.

223 Jamieson GA, Greenwalt TJ, eds. Blood Substitutes and Plasma Expanders. New York: Alan R Liss Inc., 1978:340.

224 Jeong ST, Byun SM. Decreased agglutinability of methoxy–polyethylene glycol attached red blood cells: significance as a blood substitute. Artificial Cells, Blood Substitutes and Immobilization Biotechnology, an International Journal 1996; 24:503–511.

225 Jia L, Bonaventura C, Bonaventrua J, Stamler JS. S–nitrosohemoglobin: a dynamic activity of blood involved in vascular control. Nature 1996; 380:221–226.

226 Jing M, Ledvina MA, Bina S, Hart JL, Muldoon SM. Effects of halogenated and non–halogenated anesthetics on diaspirin crosslinked hemoglobin™–induced contractions of porcine pulmonary veins. Artificial Cells, Blood Substitutes and Immobilization Biotechnology, an International Journal 1995; 23:487–494.

227 Jing M, Panico FG, Panico JL, Ledvina MA, Bina S, Muldoon SM. Diaspirin crosslinked hemo-globin does not alter isolated human umbilical artery or vein tone. Artificial Cells, Blood Substi-tutes and Immobilization Biotechnology, an International Journal 1996; 24:621–628.

228 Joyner, MJ, Faust RJ. Blood Substitutes: What is the target? In: RM Winslow, KD Vandegriff, M Intaglietta, eds. Blood Susbtitutes: New Challenges. Boston: Birkhäuser, 1996:15–33.

229 Jweier JL et al. Abstract volume for VI International Symposium on Blood Substitutes. Artificial Cells, Blood Substitutes and Immobilization Biotechnology, an International Journal 1996; Vol 24, issue 4.

230 RI Roth, Levin J. Hemoglobin, a newly recognized lipopolysaccharide (LPS)–binding protein that enhances LPS biological activity. J Biol Chem 1994; 269:25078–25084.

231 Kan P, Chen W–K, Lee CJ. Simulation of oxygen saturation of hemoglobin solution, RBC sus-pension and hemosome by a neural network system. Artificial Cells, Blood Substitutes and Im-mobilization Biotechnology, an International Journal 1996; 24:143–151.

232 Kasper SM, Walter M, Grune F, Bischoff A, Erasmi H, Buzello W. Effects of hemoglobin–based oxygen carrier (HBOC–201) on hemodynamics and oxygen transport in patients undergoing preoperative hemodilution for efective abdominal aortic surgery. Anesth Analg 1996; 83:921–927.

233 Kaufman RJ, Clinical development of perfluorocarbon–based emulsions as red cell substitutes. In: Winslow RM, Vandergriff KD, Intaglietta M, eds. Blood Substitutes: Physiological Efficacy. Boston: Birkhäuser, 1995.

234 Keipert PE, Minkowitz J, Chang TMS. Crosslinked stroma–free polyhemoglobin as a potential blood substitute. Int J Artif Organs 1982; 5(6):383.

235 Keipert PE, Chang TMS. In vivo assessment of pyridoxalated polyhemoglobin as an artificial red cell substitute in rats. Trans Am Soc Artif Intern Organs 1983; 29:329.

236 Keipert PE, Chang TMS. Preparation and in–vitro characteristics of pyridoxalated polyhemoglobin as blood substitute. Appl Biochem & Biotechnol 1984; 10:133.

237 Keipert PE, Chang TMS. Pyridoxalated polyhemoglobin as a blood substitute for resuscitation of lethal hemorrhagic shock in conscious rats. Biomat Med Dev Artif Organs 1985; 13:1–15.

238 Keipert P, Chang TMS. In–vivo effects of total and partial isovolemic exchange transfusion in fully conscious rats using pyridoxalated polyhemoglobin solution as a colloidal oxygen–delivery blood substitute. Vox Sanguinis 1987; 53:7–14.

239 Keipert PE, Chang TMS. Pyridoxalated–polyhemoglobin solution: A low viscosity oxygen–de-livering blood replacement fluid with normal oncotic pressure and long–term storage feasibility. Biomaterials, Artificial Cells and Artificial Organs 1988; 16:185–196.

240 Keipert PE, Verosky M, Triner L. Metabolism, distribution, excretion of HbXL: A non–dissociat-ing interdimerically crosslinked hemoglobin with exceptional oxygen offloading capability. Biomaterials, Artificial Cells and Artificial Organs 1988; 16:643–645.

241 Keipert PE. Use of oxygent, a perfluorochemical–based oxygen carrier, as an alternative to intraoperative blood transfusion. Artificial Cells, Blood Substitutes and Immobilization Biotech-nology, an International Journal 1995; 23:381–394.

242 Keipert PE, Faithfull NS, Roth DJ, Bradley JD, Batra S, Jochelson P, Flaim KE. Supporting tis-sue oxygenation during acute surgical bleeding using a perfluorochemical–based oxygen carrier. Adv Exp Med Biol 1996; 388:603–609.

243 Kim HW, Chen F, Greenburg AG. Hemoglobin–based double (exchange transfusion – carbon clear-ance) model for testing post–resuscitation reticuloendothelial function. Biomaterials, Artificial Cells and Immobilization Biotechnology 1992; 20:777–780.

244 Kim HW, Greenburg AG. Hemoglobin mediated vasoactivity in isolated vascular rings. Artificial Cells, Blood Substitutes and Immobilization Biotechnology, an International Journal 1995; 23:303–309.

245 Kim HW, Greenberg G. Abstract volume for VI International Symposium on Blood Substitutes. Artificial Cells, Blood Substitutes and Immobilization Biotechnology, an International Journal 1996; Vol 24, issue 4.

246 Kim HW, Greenburg AG. Ferrous hemoglobin scavenging of endothelium derived nitric oxide is a principal mechanism for hemoglobin mediated vasoactivities in isolated rat thoracic aorta trans-

fusion. Artificial Cells, Blood Substitutes and Immobilization Biotechnology, an International Journal 1997; 25:121–134.

247 Kim HW, Breiding P, Greenburg AG. Enhanced modulation of hypotension in endotoxemia by concomitant nitric oxide synthesis inhibition and nitric oxide scavenging. Artificial Cells, Blood Substitutes and Immobilization Biotechnology, an International Journal 1997; 25:153–162.

248 Kim MS, Kim HW, Sweeney JD, Greenburg AG. Decreased whole blood factor ix activity following hemodilution with purified hemoglobin in–vitro. Artificial Cells, Blood Substitutes and Immobilization Biotechnology, an International Journal 1997; (in press).

249 Klein HG. Oxygen carriers and transfusion medicine. Artificial Cells, Blood Substitutes and Immobilization Biotechnology, an International Journal 1994; 22:123–135.

250 Kluger R, Jones RT, Shih DT. Crosslinking hemoglobin by design: lessons from using molecular clamps. Artificial Cells, Blood Substitutes and Immobilization Biotechnology, an International Journal 1994; 22:415–428.

251 Kochetygov NI, Gorkoun AV, Gerbut KA, Sedova LA, Mikhailova LG. Study of therapeutic efficiency of new blood substitutes in experimental hemorrhagic shock. Artificial Cells, Blood Substitutes and Immobilization Biotechnology, an International Journal 1996; 24:685–691.

252 Kreimeier U, Messamer K. Hemodilution in clinical surgery: state of the art 1996. World J Surg 1996; 20:1208–1217.

253 Krishnamurti C, Alving B. Biological consequences of crosslinked hemoglobin in animal models of surgery and endotoxemia. In: Winslow RM, Vandegriff KD, Intaglietta M, eds. Blood Sunstitutes: New Challenges. Boston: Birkhäuser, 1996:99–111.

254 Kumar R. Recombinant hemoglobins as blood substitutes: a biotechnology perspective. Proc Soc Exp Biol Med 1995; 208:150–158.

255 Kumar A, Sen AP, Saxena PR, Gulati A. Resuscitation with diaspirin crosslinked hemoglobin increases cerebral and renal blood perfusion in hemorrhaged rats. Artificial Cells, Blood Substitutes and Immobilization Biotechnology, an International Journal 1997; 25:85–94.

256 Lalla FR, Ning J, Chang TMS. Effects of pyridoxalated polyhemoglobin and stroma–free hemoglobin on ADP–induced platelet aggregation. Biomaterials, Artificial Cells and Artificial Organs 1989; 17:271–290.

257 Langermans JA, van Vuren–van der Huslt M, Bleeker WK. Safety evaluation of a polymerized hemoglobin solution in a murine infection model. J Lab Clin Med 1996; 127:428–434.

258 Lee R, Neya K, Svizzero TA, Vlahakes GJ. Limitations of the efficacy of hemoglobin–based oxygen–carrying solutions. J Appl Physiol 1995; 79:236–242.

259 Lemon DD, Dohertym DH, Curry SR, Mathews AJ, Doyle MP, Fattor TJ, Olson JS. Abstract volume for VI International Symposium on Blood Substitutes. Artificial Cells, Blood Substitutes and Immobilization Biotechnology, an International Journal 1996; Vol 24, issue 4.

260 Leppaniemi A, Soltero R, Burris D, Pikoulis E, Ratigan J, Waasdorp C, Hufnagel H, Malcolm D. Early resuscitation with low–volume PolyDCLHB is effective in the treatment of shocked induced by penetrating vascular injury. J Trauma 1996; 40:242–248.

261 Leslie SB, Puvvada S, Ratna BR, Rudolph AS. Encapsulation of hemoglobin in a bicontinuous cubic phase lipid. Biochim Biophys Acata 1996; 1285:246–254.

262 Levin J, Noth RI, Kaca W, Yoshida M, Su D. Hemoglobin–endotoxin interactions. In: Winslow RM, Vandegriff KD, Intaglietta M, eds. Blood Sunstitutes: New Challenges. Boston: Birkhäuser, 1996:185–202.

263 Lowe KC, Antony P, Wardrop J, Davey MR, Power JB. Abstract volume for VI International Symposium on Blood Substitutes. Artificial Cells, Blood Substitutes and Immobilization Biotechnology, an International Journal 1996; Vol 24, issue 4.

264 Looker D, Durfee S, Shoemaker S, Mathews A, Nagai Kiyoshi, Stetler G. Production of recombinant hemoglobin specifically engineered to enhance delivery and circulating half–life: A recombinant cell–free blood substitute. Biomaterials, Artificial Cells and Immobilization Biotechnology 1991; 19(2):418.

265 Looker D, Abbott–Brown D, Cozart P et al. A human recombinant hemoglobin designed for use as a blood substitute. Nature 1992; 356:258–260.

---

266 Lowe KC, Furmidge BA, Thomas S. Haemolytic properties of pluronic surfactants and effects of purification. Artificial Cells, Blood Substitutes and Immobilization Biotechnology, an International Journal 1995; 23:135–139.

267 Lowe KC, Anthony P, Davey MR. Enhanced protoplast growth at the interface between oxygenated fluorocarbon liquid and aqueous culture medium supplemented with pluronic F–68. Artificial Cells, Blood Substitutes and Immobilization Biotechnology, an International Journal 1995; 23:417–422.

268 Lowe KC, Anthony P, Wardrop J, Davey, Power JB. Perfluorochemicals and cell biotechnology. Artificial Cells, Blood Substitutes and Immobilization Biotechnology, an International Journal 1997; (in press).

269 Lutz J, Kettemann M, Racz I, Noth U. Several methods utilized for the assessment of biocompatibility of perfluorochemicals. Artificial Cells, Blood Substitutes and Immobilization Biotechnology, an International Journal 1995; 23:407–415.

270 Lutz J, Kettemann M, Bruch E, Barnikol WKR, Pux B. Abstract volume for VI International Symposium on Blood Substitutes. Magnin A, speaker. Artificial Cells, Blood Substitutes and Immobilization Biotechnology, an International Journal 1996; Vol 24, issue 4.

271 Mallick A, Bodenham AR. Modified haemoglobins as oxygen transporting blood substitutes. Br J Hosp Med 1996; 55:443–448.

272 Malcolm D, Kissinger D, Garrioch M. Diaspirin crosslinked hemoglobin solution as a resuscitative fluid following severe hemorrhage in the rat. Biomaterials, Artificial Cells and Immobilization Biotechnology 1992; 20:495–498.

273 Marks DH, Patressi J, Chaudry IT. Effects of pyridoxalated stabilized stroma–free hemoglobin solution on the clearance of intravascular lipid by the reticuloendothelial system. Circ Shock 1985; 16:165–172.

274 Manjula BN, Smith PK, Malavalli A, Acharya AS. Intramolecular crosslinking of oxy hemoglobin by bis sulfosuccinimidyl suberate and sebacate: generation of crosslinked hemoglobin with reduced oxygen affinity. Artificial Cells, Blood Substitutes and Immobilization Biotechnology, an International Journal 1995; 23:311–318.

275 Manning JM. Examples of chemical modification and recombinant DNA approaches with hemoglobin. Transfus Clin Biol 1996; 3:109–111.

276 Manning JM. Random Chemical modification of hemoglobin to identify chloride binding sites in the central dyad axis: their role in control of oxygen affinity. Artificial Cells, Blood Substitutes and Immobilization Biotechnology (Winslow R, guest editor) 1994; 22:199–206.

277 Marchand G, Dunlap E, Farrell L, Nigro C, Burhop K. Resuscitation with increasing doses of diaspirin crosslinked hemoglobin in swine. Artificial Cells, Blood Substitutes and Immobilization Biotechnology, an International Journal 1996; 24:469–487.

278 Marshall T, Weltzer J, Hai T, Estep T, Farmer M. Trace element analysis in diaspirin crosslinked hemoglobin solutions. Biomaterials, Artificial Cells and Immobilization Biotechnology 1992; 20:453–456.

279 Mathews AJ, Durfee SL, Looker DL, Trimble SP, Cozart PE, Stetler GL. Functional properties of potential blood substitutes: Protein engineering of human hemoglobin. Biomaterials, Artificial Cells and Immobilization Biotechnology 1991; 19:431.

280 Matsumura H, Araki H, Morioka T, Nishi K. Pyridoxalated hemoglobin polyoxyethylene conjugate (PHP) on the endothelium–dependent relaxation in rat mesemteric arterioles. Biomaterials, Artificial Cells and Immobilization Biotechnology 1992; 20:679–682.

281 Matsushita M, Yabuki A, Malchesky PS, Harasaki H, Nose Y. In vivo evaluation of a pyridoxalated–hemoglobin–poloxyethylene conjugate. Bimoaterials, Artificial Cells and Artificial Organs 1988; 16:247–260.

282 Matsushita S, Sakakibara Y, Jikuya T, Atsumi N, Tsutui T, Okamura K, Ijima H, Mitsui T, Hori M. Effect of stabilized hemoglobin as a component of cardioplegia on warm ischemic heart. J Biomaterials, Artificial Cells and Immobilization Biotechnology 1992; 20:703–708.

283 McCall WG. Physiological and practical considerations of fluid management. CRNA 1996; 7:62-70.

284 Mackay Z, Shugufta Q, Din M, Guru AA. Hemodilution in complicated high velocity vascular injuries of limbs. J Cardiovasc Surg (Torino) 1996; 37:217–221.

285 McDonagh PF, Wilson DS. The initial response of blood leukocytes to incubation with perfluorocarbon blood substitute emulsions. Artificial Cells, Blood Substitutes and Immobilization Biotechnology, an International Journal 1995; 23:439–447.

286 McKenzie JE, Scandling DM, West SD, Linz PE, McLuckie AE, Savage RW. Abstract volume for VI International Symposium on Blood Substitutes. Artificial Cells, Blood Substitutes and Immobilization Biotechnology, an International Journal 1996; Vol 24, issue 4.

287 Meinert H. Abstract volume for VI International Symposium on Blood Substitutes. Artificial Cells, Blood Substitutes and Immobilization Biotechnology, an International Journal 1996; Vol 24, issue 4.

288 Menu P, Faivre B, Labrude P, Grandgeorge M, Vigneron C. Possible importance of chromatographic purification position in a blood substitute elaboration process. Biomaterials, Artificial Cells and Immobilization Biotechnology 1992; 20:443–446.

289 Menu P, Donner M, Faivre B, Labrude P, Vigneron C. In vitro effect of dextran–benzene–tetra–carboxylate hemoglobin on human blood rheological properties. Artificial Cells, Blood Substitutes and Immobilization Biotechnology, an International Journal 1995; 23:319–330.

290 Messmer K. Characteristics, effects and side effects of plasma substitutes. In: Lowe KC, ed. Blood Substitutes. Preparation, Physiology and Medical Applications 1988:51–70.

291 Miles PJ, Langley KV, Stacey CJ, Talarico TL. Detection of residual polyethylene glycol derivatives in pyridoxalated–hemoglobin–polyoxyethylene conjugate. Artificial Cells, Blood Substitutes and Immobilization Biotechnology, an International Journal 1997; (in press).

292 Miller IF, Mayoral J, Djordjevich L, Kashani A. Hemodynamic effects of exchange transfusions with liposome–encapsulated hemoglobin. Biomaterials, Artificial Cells and Artificial Organs 1988; 16:281–288.

293 Minato N, Sasaki T, Sakuma I, Shiono M, Takatani S, Nose Y. Potential clinical applications of the oxygen carrying solutions. J. Biomaterials, Artificial Cells and Immobilization Biotechnology 1992; 20:221–232.

294 Mitsuno T, Ohyanagi. Present status of clinical studies of fluosol–DA (20%) in Japan. In: Tremper KK, ed. Perfluorochemical Oxygen Transport. Boston: Little Brown & Co, 1985:169–184.

295 Mitsuno T, Naito R, eds. Perfluorochemical Blood Substitutes. Amsterdam: Excerpta Medica, 1979.

296 Mobed M, Chang TMS. Preparation and surface characterization of carboxymethylchitus incorporated submicron bilayer lipid membrane artificial cells (liposomes) encapsulating hemoglobin. Biomaterials, Artificial Cells and Immobilization Biotechnology 1991; 19:731–744.

297 Mobed M, Chang TMS. Purification and quantitative determination of carboxymethylchitin–incorporated into submicron bilayer–lipid membrane artificial cells (liposomes) encapsulating hemoglobin. Biomaterials, Artificial Cells and Immobilization Biotechnology, 1992; 20:329–336.

298 Mobed M, Chang TMS. Surface characterization of carboxymethychitin–incorporated liposomes encapsulating hemoglobin. Biomaterials, Artificial Cells and Immobilization Biotechnology, 1992; 20:369–376.

299 Mobed M, Nishsiya T, Chang TMS. Preparation of carboxymethylchitin–incorporated submicron bilayer–lipid membrane artificial cells (liposomes) encapsulating hemoglobin. Biomaterials, Artificial Cells and Immobilization Biotechnology 1992; 20:325–328.

300 Mobed M, Nishiya T, Chang TMS. Preparation of carboxymethylchitin–incorporated liposomes encapsulating hemoglobin. Biomaterials, Artificial Cells and Immobilization Biotechnology 1992; 20:365–368.

301 Mobed M, Chang TMS. The importance of standardization of carboxymethylchitin concentration by the dye–binding capacity of alcian blue. Artificial Cells, Blood Substitutes and Immobilization Biotechnology, an International Journal 1996; 24:107–120.

302 Mok W, Chen DE, Mazur A. Covalent linkage of subunits of hemoglobin. Fed Proc 1975; 34:1458.

303 Morton P, Ilan E, Chang TMS. Divinyl sulfone modified uncrosslinked bovine hemoglobin with improved P_{50} at low temperatures: A potential oxygen carrier for organ preservation. Biomaterials, Artificial Cells and Immobilization Biotechnology 1991; 19:446–.

---

304 Moss GS, DeWoskin R, Rosen AL, Levine H, Palani CK. Transport of oxygen and carbon dioxide by hemoglobin–saline solution in the red cell–free primate. Surg Gynecol Obstet 1976; 142:357.

305 Moss GS, Gould SA, Sehgal LR, Sehgal HL, Rosen AL. Hemoglobin solution – from tetramer to polymer. Biomaterials, Artificial Cells and Artificial Organs 1988; 16:57–69.

306 Motterlini R, Foresti R, Vandegriff K, Winslow RM. The autoxidation of alpha–alpha crosslinked hemoglobin: A possible role in the oxidative stress to endothelium. Artificial Cells, Blood Substitutes and Immobilization Biotechnology, an International Journal 1995; 23:291–301.

307 Mouelle P, Labrude P, Grandgeorge M, Vigneron C. A temporary blood substitute based on dextran hemoglobin conjugates. III Effects on Guinea–pig heart perfusion and myocardial ischemia reperfusion. Biomaterials, Artificial Cells and Immobilization Biotechnology 1992; 20:697–702.

308 Muldoon S, Hart J, Ledvina M. Abstract volume for VI International Symposium on Blood Substitutes. Artificial Cells, Blood Substitutes and Immobilization Biotechnology, an International Journal 1996; Vol 24, issue 4.

309 Mullett CJ, Polak MJ. Vasoreactivity of Fluosol™ perfluorocarbon emulsion vs Earl's balanced salt/albumin perfusates in the isolated perfused rat lung. Artificial Cells, Blood Substitutes and Immobilization Biotechnology, an International Journal 1995; 23:449–457.

310 Murray JA, Ledlow A, Launspach J et al. The effects of recombinant human hemoglobin on esophageal motor functions in humans. Gastroenterology 1995; 109:1241–1248.

311 Myhre BA. The first recorded blood transfusions: 1656 to 1668. Transfusion 1990; 30:358.

312 Naito R, Yokoyama K. An improved perfluorodecalin emulsion. In: GA Jamieson, TJ Greenwalt, eds. Blood Substitutes and Plasma Expanders. New York: Alan R Liss Inc, 1978:81.

313 Nakai K, Abe A, Matsuda N, Kobayashi M, Ikeda H, Sekiguchi S, Tsuchida E. Development of analytical methods to evaluate SFH. Biomaterials, Artificial Cells and Immobilization Biotechnology, 1992; 20:446–452.

314 Nho K, Glower D, Bredehoeft S, Shankar H, Shorr R, Abuchowski A. PEG–bovine hemoglobin: Safety in a canine dehydrated hypovolemic–hemorrhagic shock model. Biomaterials, Artificial Cells and Immobilization Biotechnology 1992; 20:511–524.

315 Nho K, Linberg R, Johnson M, Gilbert C, Shorr R. PEG–Hemoglobin: An efficient oxygen–delivery system in the rat exchange transfusion and hypovolemic shock models. Artificial Cells, Blood Substitutes and Immobilization Biotechnology 1996; 22:795–803.

316 Ni Y, Klein DH, Song D. Recent development in pharmacokinetic modeling of perfluorocarbon emulsion. Artificial Cells, Blood Substitutes and Immobilization Biotechnology, an International Journal 1996; 24:81–90.

317 Ning J, Chang TMS. Effects of homologous and heterologous stroma–free hemoglobin and polyhemoglobin on blood cell counts, complement activation and coagulation factors in rats. Biomaterials, Artificial Cells and Artificial Organs 1987; 15:380.

318 Ning J, Chang TMS. Effects of stoma–free Hb and polyhemoglobin on complement activation blood cell counts and coagulation factors in rats. Biomaterials, Artificial Cells and Artificial Organs 1988; 16:651–652.

319 Ning J, Chang TMS. In–vivo effects of stroma–free hemoglobin and polyhemoglobin on coagulation factors in rats. Int J Artificial Organs 1990; 13:509–516.

320 Ning J, Chang TMS. Effects of homologous and heterologous stroma–free hemoglobin and polyhemoglobin on complement activation, leukocytes and platelets. Biomaterials, Artificial Cells and Artificial Organs 1990; 18:219–233.

321 Ning J, Anderson PJ, Biro GP. Resuscitation of bled dogs with pyridoxalated–polymerized hemoglobin solution. Biomaterials, Artificial Cells and Immobilization Biotechnology 1992; 20:525–530.

322 Ning J, Chang TMS. Measurement of complement activation by CH50 in rats. Biomaterials, Artificial Cells and Artificial Organs 18:203–218.

323 Nishiya T, Dasgupta M, Okumura Y, Chang TMS. Circular dichroism study of membrane dynamics focused on effect of monosialoganglioside. J Biochemistry 1988; 104:62–65.

324 Nishiya T, Lam RTT, Eng F, Zerey M, Lau S. Mechanistic study on toxicity of positively charged liposomes containing stearylamine to blood. Artificial Cells, Blood Substitutes and Immobilization Biotechnology, an International Journal 1995; 23:505–512.

325 Nishiya T, Jain B. Study on in vitro stability of polymerized liposomes. Artificial Cells, Blood Substitutes and Immobilization Biotechnology, an International Journal 1996; 24:43–50.

326 Noth U, Morrissey SP, Deichmann R, Jung S, Adolf H, Haase A, Lutz J. Perfluro–15–crown–5–ether labelled macrophages in adoptive transfer experimental allergic encephalomyelitis. Artificial Cells, Blood Substitutes and Immobilization Biotechnology, an International Journal 1997; (in press).

327 O'Donnell JK, Swanson M, Pilder S, Martin M, Hoover K, Huntress V, Karet C, Pinkert C, Lago W, Logan J. Production of human hemoglobin in transgenic swine. Biomaterials, Artificial Cells and Immobilization Biotechnology 1992; 20:149–.

328 Ogata Y, Goto H, Kimura T, Fukui H. Abstract volume for VI International Symposium on Blood Substitutes. Artificial Cells, Blood Substitutes and Immobilization Biotechnology, an International Journal 1996; Vol 24, issue 4.

329 Ogden JE, Woodrow J, Perks K, Harris R, Coghlan D, Wilson MT. Expression and assembly of functional human hemoglobin in SACCHAROMYCES CEREVISIAE. Biomaterials, Artificial Cells and Immobilization Biotechnology 1992; 20:473–477.

330 Ogden JE, Parry ES. The development of hemoglobin solutions as red blood cell substitutes. Int Anesthesiol Clin 1995; 33:115–129.

331 Ohyanagi H, Toshima K, Sekita M, Okamoto M, Itoh T, Mitsuno T, Naito R, Suyama T, Yokoyama K. Clinical study of perfluorochemical whole blood substitute. Clinical Therapeutics 1979; 2:306.

332 Olson JS. Genetic engineering of myoglobin as a simple prototype for hemoglobin–based blood substitutes. Artificial Cells, Blood Substitutes and Immobilization Biotechnology 1994; 22:429–442.

333 Olson JS. Abstract volume for VI International Symposium on Blood Substitutes. Artificial Cells, Blood Substitutes and Immobilization Biotechnology, an International Journal 1996; Vol 24, issue 4.

334 Olson JS, Eich RF, Smith LP, Warren JJ, Knowles BC. Protein engineering strategies for designing more stable hemoglobin–based blood substitutes. Artificial Cells, Blood Substitutes and Immobilization Biotechnology, an International Journal 1997; 25:227–241.

335 Panter SS, Vandegriff KD, van PO, Egan RF. Assessment of hemoglobin–dependent neurotoxicity: alpha–alpha cross–lined hemoglobin. Artificial Cells, Blood Substitutes and Immobilization Biotechnology 1994; 22:687–694.

336 Pavlik PA, Boyd MK, Olsen KW. Molecualar Dynamics of a hemoglobin crosslinking reaction. Biopolymers 1996; 39:615–618.

337 Payne JW. Glutaraldehyde crosslinked protein to form soluble molecular weight markers. Biochem J 1973; 135:866–873.

338 Peerless SJ, Nakamura R, Rodriguez–Salazar A, Hunter IG. Modification of cerebral ischemia with fluosol. Stroke 1985; 16:38.

339 Perutz MF. Stereochemical mechanism of oxygen transport by hemoglobin. Proc R Soc Lond B 1980; 208:135.

340 Perutz MF. Myoglobin and hemoglobin: Role of distal residues in reactions with haem ligands. Trends Biochem Science 1989; 14:42–44.

341 Pickelmann S, Nolte D, Messmer K. Abstract volume for VI International Symposium on Blood Substitutes. Artificial Cells, Blood Substitutes and Immobilization Biotechnology, an International Journal 1996; Vol 24, issue 4.

342 Phillips WT, Rudolph AS, Klipper R. Biodistribution studies of liposome encapsulated hemoglobin (LEH) studied with a newly developed 99m–technetium liposome label. Biomaterials, Artificial Cells and Immobilization Biotechnology 1992; 20:757–760.

343 Phillips R, Mawhinney T, Harmata M, Smith D. Characterization of gallus domesticus–n–acetyl–galactosaminidase blood group A_2 activity. Artificial Cells, Blood Substitutes and Immobilization Biotechnology, an International Journal 1995; 23:63–79.

344 Phiri JB, Senthil V, Grossweiner LI. Diffuse optics determination of hemoglobin derivatives in red blood cells and liposome encapsulated hemoglobin. Artificial Cells, Blood Substitutes and Immobilization Biotechnology, an International Journal 1995; 23:23–38.

References

345 Pliura DH, Wiffen DE, Ashraf SS et al. Purification of hemoglobin by displacement chromatography. U.S. Patent 5,545,328, August 13, 1996.

346 Poli de Figueiredo LF, Elgjo GI, Mathru M, Rocha e Silva M, Kramer GC. Hypertonic acetate–aahemoglobin for small volume resuscitation of hemorrhagic shock. Artificial Cells, Blood Substitutes and Immobilization Biotechnology, an International Journal 1997; 25:61–74.

347 Powell CC, Schultz SC, Malcolm DS. Diaspirin crosslinked hemoglobin (DCLHB™): More effective than lactated Ringer's solution in restoring central venous oxygen saturation after hemorrhagic shock in rats. Artificial Cells, Blood Substitutes and Immobilization Biotechnology, an International Journal 1996; 24:197–200.

348 Poznansky MJ, Chang TMS. Comparison of the enzyme kinetics and immunological properties of catalase immobilized by microencapsulation and catalase in free solution for enzyme replacement. Biochim Biophys Acta 1974; 334:103–115.

349 Pristoupil TI, Sterbikova J, Vrana M, Havlickova J, Matejckova M, Schejbalova S, Eserova L. Heart protection by cardioplegic solutions containing oxyhemoglobin pretreated by carbontetrahloride and freeze–drying with sucrose. Biomaterials, Artificial Cells and Immobilization Biotechnology 1992; 20:709–720.

350 Przybelsk R, Kisicki J, Daily E, Bounds M, Mattia–Goldberg C (Baxter Healthcare Co.). Diaspirin Crosslinked hemoglobin (DCLHb) phase I clinical safety assessment in normal healthy volunteers. #H16 Abstract, Vth International Symposium on Blood Substitutes, San Diego, March 17–20, 1993.

351 Przybelski R, Blue J, Nanvaty M, Goldberg C, Estep T, Schmitz T. Clinical studies with diaspirin crosslinked hemoglobin solution (DCLHb™): a review and update. Artficial Cells, Blood Substitutes and Immbolization Biotechnology, an International Journal 1996; 24(abstracts issue):407.

352 Przybelski R, Nanavaty J, Goldberg C, Estep T, Schmitz T. Abstract volume for VI International Symposium on Blood Substitutes. Artificial Cells, Blood Substitutes and Immobilization Biotechnology, an International Journal 1996; Vol 24, issue 4.

353 Quebec EA, Chang TMS. Superoxide dismutase and catalase crosslinked to polyhemoglobin reduces methemoglobin formation in vitro. Artificial Cells, Blood Substitutes and Immobilization Biotechnology, an International Journal 1995; 23:693–705.

354 Quebec EA, Chang TMS. Abstract volume for VI International Symposium on Blood Substitutes. Artificial Cells, Blood Substitutes and Immobilization Biotechnology, an International Journal 1996; Vol 24, issue 4.

355 Rabiner SF, Helbert JR, Lopas H, Friedman LH. Evaluation of stroma free hemoglobin solution as a plasma expander. J Exp Med 1967; 126:1127.

356 Rabinovici R, Neville LF, Rudolph AS, Feurstein G. Hemoglobin–based oxygen–carrying resuscitation fluids [editorial comments]. Crit Care Med 1995; 23:801–804.

357 Razack S, D'Agnillo F, Chang TMS. Abstract volume for VI International Symposium on Blood Substitutes. Artificial Cells, Blood Substitutes and Immobilization Biotechnology, an International Journal 1996; Vol 24, issue 4.

358 Razack S, D'Agnillo F, Chang TMS. Crosslinked hemoglobin–superoxide dismutase–catalase scavenges free radicals in a rat model of intestinal ischemia–reperfusion injury. Artificial Cells, Blood Substitutes and Immobilization Biotechnology, an International Journal 1997; 25:181–192.

359 Rhea et al. Crit Care Med 1996; 24:A3.

360 Riess JG. Fluorocabron–based in vivo oxygen transport and delivery systems. Vox Sang 1991; 61:225–239.

361 Riess J, guest editor. Blood substitutes and related products: The fluorocabon approach. Artificial Cells, Blood Substitutes and Immobilization Biotechnology, an International Journal 1994; 22:945–1511.

362 Riess, JG, Krafft MP. Abstract volume for VI International Symposium on Blood Substitutes. Artificial Cells, Blood Substitutes and Immobilization Biotechnology, an International Journal 1996; Vol 24, issue 4.

363 Riess JG, Krafft MP. Advanced fluorocarbon–based systems for oxygen and drug delivery, diagnosis. Artificial Cells, Blood Substitutes and Immobilization Biotechnology, an International Journal 1997; 25:43–52.

364 Rioux F, Drapeau C, Marceau F. Recombinant human hemoglobin (rHb1.1) selectively inhibits vasorelaxation elicied by nitric oxide donors in rabbit isolated aortic rings. J Cardiovasc Pharmacol 1995; 25:587–594.

365 Robinson MF, Dupuis NP, Kusumoto T, Liu F, Menon K, Teicher BA. Increased tumor oxygenation and radiation sensitivity in two rat tumors by a hemoglobin–based oxygen carrying preparation. Artificial Cells, Blood Substitutes and Immobilization Biotechnology, an International Journal 1995; 23:431–438.

366 Rockwell S, Kelley M, Mattrey R. Preclinical evaluation of oxygent J as an adjunct to radiotherapy. Biomaterials, Artificial Cells and Immobilization Biotechnology 1992; 20:883–893.

367 Rohlfs RJ, Vandergriff KD. Non–phospholipid liposomes: A novel method for the preparation of hemoglobin–containing lipid vesicles. In: RM Winslow, KD Vandegriff, M Intaglietta, eds. Blood Substitutes: New Challenges. Boston: Birkhäuser, 1996:163–184.

368 Rollwagen FM, Gafney WC, Pacheo ND, Davis TA, Hickey TM, Nielson TB, Rudolph AS. Multiple responses to admistration of liposome–encapsulaed hemoglobin (LEH): Effects on hematopoiesis and serum IL–6 levels. Exp Hematol 1996; 24:429–436.

369 Rosental AM, Chang TMS. The incorporation of lipid and Na^+–K^+–ATPase into the membranes of semipermeable microcapsules. J Membrane Sciences 1980; 6:329–338.

370 Roth RI, Kaca W. Toxicity of hemoglobin solutions: hemoglobin is a lipopolysacchride (LPS) binding protein which enhances LPS biological activity. Artificial Cells, Blood Substitutes and Immobilization Biotechnology (Winslow R, guest editor), 1994; 22:387–398.

371 Roth RI, Levin J, Chapman KW et al. Production of modified crosslinked cell–free hemoglobin for human use: the role of quantitative determination of endotoxin contamination. Transfusikon 1933; 33:919–924.

372 Rudolph AS. Encapsulated hemolgobin: Current issues and future goals. Artificial Cells, Blood Substitutes and Immobilization Biotechnology, an International Journal 1994; 22:347–360.

373 Rudloph AS. Encapsulation of hemoglobin in liposomes. In: Winslow RM, Vandergriff KD, Intaglietta M, eds. Blood Substitutes: Physiological Basis of Efficacy. Boston: Bikhäuser, 1995.

374 Rudolph AS, Sulpizio T, Kwasiborski V, Cliff R, Rabinovici R, Feuerstein G. Abstract volume for VI International Symposium on Blood Substitutes. Artificial Cells, Blood Substitutes and Immobilization Biotechnology, an International Journal 1996; Vol 24, issue 4.

375 Runge TM, McGinity JW, Frisbee SE, Briceno JC, Ottmers SE, Calhoon JH, Hantler CB, Korvick DL, Ybarra JR. Enhancement of Brain pO2 During Cardiopulmonary Bypass Using a Hyperosmolar Oxygen Carrying Solution. Artificial Cells, Blood Substitutes and Immobilization Biotechnology, an International Journal 1997; (in press).

376 Satzler RK, Arfors KE, Tuma R, Ma L, Timble CE, Hsia CJC, Lehr HA. Abstract volume for VI International Symposium on Blood Substitutes. Artificial Cells, Blood Substitutes and Immobilization Biotechnology, an International Journal 1996; Vol 24, issue 4.

377 Sakai H, Hamada K, Takeoeka S, Nishide H, Tsuchida E. Physical properties of hemoglobin vesicles as red cell substitutes. Biotechnol Prog 1996; 12:119–125.

378 Sanders KE, Ackers G, Sligar S. Enginnering and desugn of blood substitues. Curr Opin Struct Biol 1996; 6:534–540.

379 Savitsky JP, Doozi J, Black J, Arnold JD. A clinical safety trial of stroma free hemoglobin. Clin Pharm Ther 1978; 23:73.

380 Schrieber GB, Busch MP, Kleinman SH, Korelitz JJ. The risk of transfusion transmitted viral infections. N Engl J Med 1996; 334:1685–1690.

381 Schultz SC, Powell CC, Burris DG et al. The efficacy of diaspirin crosslinked hemoglobin solution resuscitation in a model uncontrolled haemorrhage. J Trauma 1994; 37:408–412.

382 Schultz SC, Powell CC, Bernard E, Malcolm DS. Diaspirin crosslinked hemoglobin (DCLHb) attenuates bacterial translocation in rats. Artificial Cells, Blood Substitutes and Immobilization Biotechnology, an International Journal 1995; 23:647–664.

383 Sehgal LR, Rosen AL, Gould SA, Sehgal HL, Dalton L, Mayoral J, Moss GS. In vitro and in vivo characteristics of polymerized pyridoxalated hemoglobin solution. Fed Proc 1980; 39:2383.

384 Sehgal LR, Rosen AL, Gould SA, Sehgal HL, Moss GS. Preparation and in vitro characteristics of polymerized pyridoxalated hemoglobin. Transfusion 1983; 23:158.

385 Sehgal LR, Gould SA, Rosen AL, Sehgal HL, Moss GS. Polymerized pyridoxalated hemoglobin: a red cell substitute with normal oxygen capacity. Surgery 1984; 95:433.

386 Sehgal LR, Sehgal HL, Rosen AL, Gould SA, De Woskin R, Moss GS. Characteristics of polymerized pyridoxalated hemoglobin. Biomaterials, Artificial Cells and Artificial Organs 1988; 16:173–183.

387 Sekiguchi S, Ito K, Kobayash M, Ototake N, Kosuda M, Kwon KW, Ikeda H. Preparation of virus–free pyridoxalated hemoglobin from the blood of HBV or HTLV–I healthy carriers. Biomaterials, Artificial Cells and Artificial Organs 1988; 16:113–121.

388 Sekiguchi S, ed. Transfusion and Hematopoietic Stem Cells. Oxford U.K.: Blackwell Science Publisher 1996:278.

389 Sekiguchi S. The impact of red cell substitutes on the blood service in Japan. Artificial Cells, Blood Substitutes and Immobilization Biotechnology, an International Journal 1997; 25:53–60.

390 Sekiguchi S. Studies on the quality control of stroma free hemoglobin. Biomaterials, Artificial Cells and Immobilization Biotechnology 1992; 20:407–414.

391 Sen AP, Dong Y, Gulati A. Effect of diaspirin crosslinked hemoglobin on systemic and regional blood circulation in pregnant rats. Artificial Cells, Blood Substitutes and Immobilization Biotechnology, an International Journal 1997; (in press).

392 Shah N, Mehra A. Modeling of oxygen uptake in perfluorocarbon emulsions. Some comparisons with uptake by blood. ASAIO J 1996; 42:181–189.

393 Sherwood RL, McCormick DL, Zheng S, Beissinger RL. Influence of steric stabilization of liposome–encapsulated hemoglobin on listeria monocytogenes host defence. Artificial Cells, Blood Substitutes and Immobilization Biotechnology, an International Journal 1995; 23:665–679.

394 Shih DTB, Noboru N, Dickey B, Jones RT, Nagai K. Function and stability of low–affinity variant and engineered hemoglobins. Biomaterials, Artificial Cells and Immobilization Biotechnology 1991; 19:483.

395 Shoemaker S, Stetler G, Looker D, Bates–Hill R, Abbott–Brown D. Recombinant human hemoglobin has been engineered to decrease toxicity and improve efficacy. Biomaterials, Artificial Cells and Immobilization Biotechnology 1991; 19:484–489.

396 Shoemaker S, Gerber M, Evans G, Paik L, Scoggin C (Somatogen Co. USA). Initial clinical experience with a rationally designed genetically engineered recombinant human hemoglobin. Artificial Cells, Blood Substitutes and Immobilization Biotechnology, an International Journal 1994; 22:457–465.

397 Shorr RG, Viau AT, Abuchowski A. Phase 1B safety evaluation of PEG hemoglobin as an adjuvant to radiation therapy in human cancer patients. Artficial Cells, Blood Substitutes and Immbolization Biotechnology, an International Journal 1996; 24:(abstracts issue) 407.

398 Shum KL, Leon A, Viau AT, Pilon D, Nucci M, Shorr RGL. The physiological and histopathological response of dogs to exchange transfusion with polyethylene glycol–modified bovine hemoglobin (PEG–Hb). Artificial Cells, Blood Substitutes and Immobilization Biotechnology, an International Journal 1996; 24:655–683.

399 Simoni J, Simoni G, Garcia EL, Prien SD, Tran RM, Feola M, Shires GT. Protective effect of selenium on hemoglobin mediated lipid peroxidation in vivo. Artificial Cells, Blood Substitutes and Immobilization Biotechnology, an International Journal 1995; 23:469–486.

400 Simoni J, Simoni G, Newman G, Bartsell A, Feola M. Abstract volume for VI International Symposium on Blood Substitutes. Artificial Cells, Blood Substitutes and Immobilization Biotechnology, an International Journal 1996; Vol 24, issue 4.

401 Simoni J, Simoni G, Lox CD, Prien SD, Shires GT. Modified hemoglobin solution with desired pharmacological properties does not activate nuclear transcription factor NF–κb in human vascular endothelial cells. Artificial Cells, Blood Substitutes and Immobilization Biotechnology, an International Journal 1997; 25:193–210.

402 Simoni J, Simoni G, Lox CD, Prien SD, Tran R, Shires GT. Expression of adhesion molecules and von Willebrand factor in human coronary artery endothelial cells incubated with differently modified hemoglobin solutions. Artificial Cells, Blood Substitutes and Immobilization Biotechnology, an International Journal 1997; 25:211–226.

403 Sloviter H, Kamimoto T. Erythrocyte substitute for perfusion of brain. Nature 1967; 216:458.

404 Spence RK. Perfluorocarbons in the twenty–first century: Clinical applications as transfusion alternatives. Artificial Cells, Blood Substitutes and Immobilization Biotechnology, an International Journal 1995; 23:367–380.

405 Stabilini R, Palazzini G, Pietta GP, Pace M, Calatroni A, Raffaldoni E, Ghessi A, Aguggini G, Agoston A. A pyridoxalated polymerized hemoglobin solution as oxygen carrying substitute. Int J Artif Organs 1983; 6:319.

406 Swan et al. Am J Kidney Dis 1995; 26:918–923.

407 Szebeni J, Hauser H, Eskelson CD, Winterhalter KH. Factors influencing the In vitro stability of artificial red blood cells based on hemoglobin–containing liposomes. Biomaterials, Artificial Cells and Artificial Organs 1988; 16:301–312.

408 Szebeni J, Wassef NM, Rudolph AS, Alving CR. Complement activation by liposome–encapsulated hemoglobin in vitro: The role of endotoxin contamination. Artificial Cells, Blood Substitutes and Immobilization Biotechnology, an International Journal 1995; 23:355–363.

409 Szebeni J, Wassef N, Rudolph AS, Alving CR. Abstract volume for VI International Symposium on Blood Substitutes. Artificial Cells, Blood Substitutes and Immobilization Biotechnology, an International Journal 1996; Vol 24, issue 4.

410 Szebeni J, Wassef NM, Rudolph AS, Alving CR. Complement activation in human serum by liposome–encapsulated hemoglobin: the role of natural anti–phospholipid antibodies. Biochim Biophys Acta 1996; 1285:127–130.

411 Tai J, Kim HW, Greenburg AG. Endothelin–1 is not involved in hemoglobin associated vaso–activities. Artificial Cells, Blood Substitutes and Immobilization Biotechnology, an International Journal 1997; 25:135–140.

412 Takahashi A. Characterization of neo red cells (NRCs), their function and safety in–vivo tests. Artificial Cells, Blood Substitutes and Immobilization Biotechnology, an International Journal 1995; 23:347–354.

413 Takaori M, Fukui A. Treatment of massive hemorrhage with liposome encapsulated human hemoglobin (NRC) and hydroxyethyl starch (HES) in beagles. Artificial Cells, Blood Substitutes and Immobilization Biotechnology, an International Journal 1996; 24:643–653.

414 Takaori M, Fukui A. Abstract volume for VI International Symposium on Blood Substitutes. Artificial Cells, Blood Substitutes and Immobilization Biotechnology, an International Journal 1996; Vol 24, issue 4.

415 Takeoka S, Sakai H, Obgushi T, Nishide H, Tsuchida E. Abstract volume for VI International Symposium on Blood Substitutes. Artificial Cells, Blood Substitutes and Immobilization Biotechnology, an International Journal 1996; Vol 24, issue 4.

416 Takeoka S, Ohgushi T, Sakai H, Kose T, Nishide H, Tsuchida E. Construction of artificial methemoglobin reduction systems in Hb vesicles. Artificial Cells, Blood Substitutes and Immobilization Biotechnology, an International Journal 1997; 25:31–42.

417 Tam SC, Blumenstein J, Wong JT. Dextran hemoglobin. Proc Natl Acad USA, 1976; 73:2128.

418 Tanaka JI, Takino H, Malchesky PS. Does oxygen supply improve graft viability in liver preservation? Biomaterials, Artificial Cells and Immobilization Biotechnology 1992; 20:545–548.

419 Tani T, Chang TMS, Kodama M, Tsuchiya M. Endotoxin removed from hemoglobin solution using polymyxin–B immobilized fibre (PMX–F) followed by a new turbidometric endotoxin assay. Biomaterials, Artificial Cells and Immobilization Biotechnology 1992; 20:394–399.

420 Teicher BA. An overview on oxygen–carriers in cancer therapy. Artificial Cells, Blood Substitutes and Immobilization Biotechnology, an International Journal 1995; 23:395–405.

421 Thomas MJ. Royal College of Physicians, Edinburgh: final consensus statement. Consensus Conference on Autologous Transfusion 4–6 October, 1995. Vox Sang 1996; 70:183–184.

422 Thomas M, Matheson–Urbaitis B, Kwansa H, Bucci E, Fronticelli C. Introduction of negative charges to a crosslinked hemoglobin: Lack of effect on plasma half time. Artificial Cells, Blood Substitutes and Immobilization Biotechnology, an International Journal 1997; (in press).

423 Tomasulo P. Transfusion alternatives:impact on blood banking worldwide. In: Winslow RM, Vandergriff KD, Intaglietta M, eds. Blood Substitutes: Physiological Basis of Efficacy. Boston: Birkhauser, 1995.

References

135

424 Traylor RJ, Pearl RG. Crystalloid versus colloid versus colloid: all colloids are not created equal. Anesth Analg 1996; 83:209–212.

425 Tremper KK, ed. Perfluorochemical Oxygen Transport. Boston: Little, Brown & Co (Int. Anesthesiology Clinics), 1985.

426 Tsai SP, Wong JTF. Enhancement of erythrocyte sedimentation rate by polymerized hemoglobin. Artificial Cells, Blood Substitutes and Immobilization Biotechnology, an International Journal 1996; 24:513–523.

427 Tsai AG, Kerger H, Intaglietta M. Microvasular oxygen distribution: Effects due to free hemoglobin in plasma. In: Winslow RM, Vandergriff KD, Intaglietta M, eds. Blood Substitutes: New Challenges. Boston: Birkhäuser, 1996:124–131.

428 Tsuchida E, Nishide H, Ohno H. Liposome/heme as a totally synthetic oxygen carrier. Biomaterials, Artificial Cells and Artificial Organs 1988; 16:313–319.

429 Tsuchida E, Nishide H. Synthesis and characterization of artificial red cell (ARC). Biomaterials, Artificial Cells and Immobilization Biotechnology 1992; 20:337–354.

430 Tsuchida E. Stabilized hemoglobin vesicles. Artificial Cells, Blood Substitutes and Immobilization Biotechnology, an International Journal 1994; 22:467–479.

431 Tsuchida E. Abstract volume for VI International Symposium on Blood Substitutes. Artificial Cells, Blood Substitutes and Immobilization Biotechnology, an International Journal 1996; Vol 24, issue 4.

432 Ulatowski JA, Koehler RC, Nishikawa T, Traystman RJ, Razynska A, Kwansa H, Urbaitis B, Bucci E. Role of nitric oxide scavenging in peripheral vasoconstrictor response to crosslinked hemoglobin. Artificial Cells, Blood Substitutes and Immobilization Biotechnology, an International Journal 1995; 23:263–269.

433 Ulatowski JA, Nishikawa T, Matheson–Urbaitis N, Bucci E, Traysman RJ, Koehler RC. Regional blood flow alterations after bovine fumaryl–crosslinked hemoglobin transfusion and nitric oxide synthase inhibition. Crit Care Med 1996; 24:558–565.

434 Ulatowski JA, Asano Y, Koehler RC, Traystman RJ, Bucci E. Sustained endothelial dependent dilation in pial arterioles after crosslinked hemoglobin transfusion. Artificial Cells, Blood Substitutes and Immobilization Biotechnology, an International Journal 1997; 25:115–120.

435 Usuba A, Motoki R, Suzuki K, Miyauchi Y, Takahashi A. Study of effect of the newly developed artificial blood "neo red cells (RC)" on hemodynamics and blood gas transport in canine hemorrhagic shock. Biomaterials, Artificial Cells and Immobilization Biotechnology 1992; 20:531–538.

436 Usuba A, Motoki R, Ogata Y, Suzuki K, Kamitani T. Effect and safety of liposome encapsulated hemoglobin "neo red cells (NRC)" as a perfusate for total cardiopulmonary bypass. Artificial Cells, Blood Substitutes and Immobilization Biotechnology, an International Journal 1995; 23:337–346.

437 Usuba A, Motok R. Abstract volume for VI International Symposium on Blood Substitutes. Artificial Cells, Blood Substitutes and Immobilization Biotechnology, an International Journal 1996; Vol. 24, issue 4.

438 Vandegriff KD, Rohlfs RJ, Winslow RM. Kinetics of ligand binding to crosslinked hemoglobin. Biomaterials, Artificial Cells and Artificial Organs 1988; 16:647–649.

439 Vandegriff KD, LeTellier YC. A comparison of rates of heme exchanges; site–specifically crosslinked versus polymerized human hemoglobin. Artificial Cells, Blood Substitutes and Immobilization Biotechnology 1994; 22:443–456.

440 Vercellottii GM, Balla G, Balla J et al. Heme and the vasculature: an oxidative hazard that induces antioxidant defence in the endothelium. Artificial Cells, Blood Substitutes and Immobilization Biotechnology (Winslow R, guest editor) 1994; 22:687–694.

441 Vercellottii GM. Abstract volume for VI International Symposium on Blood Substitutes. Artificial Cells, Blood Substitutes and Immobilization Biotechnology, an International Journal 1996; Vol 24, issue 4.

442 Vogel WM, Hsia JC, Briggs LL, Ezss RC, Cassidy G, Apstein CS, Valeri R. Reduced coronary vasoconstrictor activity of hemoglobin solutions purified by ATP–agarose affinity chromatography. Life Science 1987; 41:89–93.

443 Vogel WM, Leiberthal CS, Apstein CS, Levinsky N, Valeri CR. Effects of stroma–free hemoglobin solutions on isolated perfused rabbit hearts and isolated perfused rat kidneys. Biomaterials, Artificial Cells and Artificial Organs 1988; 16:227–236.

444 Vogel WM, Cassidy G, Valeri CR. Effects of o–raffinose–polymerized human hemoglobin on coronary tone and cardiac function in isolated hearts. Biomaterials, Artificial Cells and Immobilization Biotechnology 1992; 20:673–678.

445 Wahr JA, Trouwborst RK, Spence RK et al. A pilot study of the efficacy of an oxygen carrying emulsion Oxtgent™, in patients undergoing surgical blood loss. Anesthesiology 1994; 80:A397.

446 Walder JA, Zaugg RH, Walder RY, Steele JM, Klotz IM. Diaspirins that crosslink alpha chains of hemoglobin: Bis(3,5–dibromosalicyl) succinate and bis(3,5–dibormosalicyl)fumarate. Biochemistry 1979; 18:4265–4270.

447 Wallace EL, Surgenor DM, Hao HS, Chapman RH, Churchill WH. Collection and transfusion of blood and blood components in the United States, 1989. Transfusion 1993; 33:139–44.

448 Walter SV, Chang TMS. Chronotropic effects of stroma–free hemoglobin and polyhemoglobin on cultured myocardiocytes derived from newborn rats. Biomaterials, Artificial Cells and Artificial Organs 1988; 16:701–703.

449 Walter SV, Chang TMS. Chronotropic effects of in–vivo perfusion with albumin, stroma–free hemoglobin and polyhemoglobin solutions. Biomaterials, Artificial Cells and Artificial Organs 1990; 18:283–299.

450 Wang YC, Lee CJ, Chen WK, Huang CI, Chen WF, Chen GJ, Lin SZ. Alteration of cerebral microcirculation by hemodilution with hemosome in awake rats. Artificial Cells, Blood Substitutes and Immobilization Biotechnology, an International Journal 1996; 24:35–42.

451 Winslow R, Chang TMS, eds. Red blood cell substitutes, Special issue. Biomaterials, Artificial Cells and Artificial Organs 1990; 18:133–342.

452 Winslow RM. Potential clinical applications for blood substitutes. Biomaterials, Artificial Cells and Immobilization Biotechnology 1992; 20:205–220.

453 Winslow R, guest editor. Special issue on "Blood substitutes: Modified hemoglobin." Artificial Cells, Blood Substitutes and Immobilization Biotechnology, an International Journal 1994; 22:360–944.

454 Winslow RM. Blood substitutes: a moving target. Nature Medicine 1995; 1:1212–1215.

455 Winslow RM, Vandergriff KD, Intaglietta M, eds. Blood Substitutes: Physiological Basis of Efficacy. Boston:Birkhäuser, 1995.

456 Winslow RM. Blood substitutes in development. Exp Opin Invest, 1996; 5:1443–1452.

457 Winslow RM, ed. Blood Substitutes. Ashley Publications Ltd, 1996.

458 Winslow RM. Blood substitute oxygen carriers designed for clinical applications. In: Winslow RM, Vandegriff KD, Intaglietta M, eds. Blood Substitutes: New Challenges. Boston: Birkhäuser, 1996:60–73.

459 Winslow RM, A Gonzales, Gonzales M. Physiologic Effects of Hemoglobin–Based Oxygen Carriers Compared to Red Blood Cells in Acute Hemorrhage [in press].

460 Wong JT. Rightshifted dextran–hemoglobin as blood substitute. Biomaterials, Artificial Cells and Artificial Organs 1988; 16:237–245.

461 Yu WP, Chang TMS. Submicron biodegradable polymer membrane hemoglobin nanocapsules as potential blood substitutes: a preliminary report. Artificial Cells, Blood Substitutes and Immobilization Biotechnology, an International Journal 1994; 22:889–894.

462 Yu WP, Chang TMS. Submicron biodegradable polymer membrane hemoglobin nanocapsules as potential blood substitutes: preparation and characterization. Artificial Cells, Blood Substitutes & Immobilization Biotechnology, an International Journal 1996; 24:169–184.

463 Yu YT, Chang TMS. Ultrathin lipid–polymer membrane microcapsules containing multienzymes, cofactors and substrates for multistep enzyme reactions. FEBS Letters 1981; 125:94–96.

464 Yu YT, Chang TMS. Immobilization of multienzymes and cofactors within lipid–polyamide membrane microcapsules for the multistep conversion of lipophilic and lipophobic substrates. Enzyme Microb Technol 1982; 4:327–331.

465 Yu YT, Chang TMS. Multienzymes and cofactors immobilized within lipid polyamide membrane microcapsules for sequential substrate conversion. Enzyme Engineering 1982; 6:163–164.

466 Zhao L, Smith JR, Eyer CL. Effects of a 100% perfluorooctylbromide emulsion on ischemia/ reperfusion injury following cardioplegia. Artificial Cells, Blood Substitutes and Immobilization Biotechnology, an International Journal 1995; 23:513–531.
467 Zheng Y, Olsen KW. Tris(3,5–dibromosalicyl) tricarballylate crosslinked hemoglobin: functional evaluation. Artificial Cells, Blood Substitutes and Immobilization Biotechnology, an International Journal 1996; 24:587–598.
468 Zuck TF, Riess G. Current status of injectable oxygen carriers. Crit Rev Clin Lab Sci 1994; 31:295–324.
469 Zuck TF. Difficulties in demonstrating efficacy of blood substitutes. Artificial Cells, Blood Substitutes and Immobilization Biotechnology (Winslow R, guest editor) 1994; 22:945–954.

Subject Index

Chemotherapy, 85
Circulation time
    Lipid–polymer, 13, 99, 102
    Lipid–protein, 13, 99, 102
    Polysaccharide, 13, 99, 102
    Sialic acid, 13, 99, 101–103
    Surface charge, 13, 99, 102-103
Coagulation, 49, 50, 61
Colloid osmotic pressure, 32, 35–37, 87
Complement, 2, 29, 49, 50, 60–64,
    66, 67, 76, 83–85, 103
Complement activation, 2, 29, 50,
    60–63, 66, 67, 83–85
Contaminant(s), 50, 60, 63, 64, 66, 67,
    78, 90
Cooperativity, 24, 25, 37, 38, 107
Crosslinked hemoglobin, 1, 3, 14,
    16–19, 21, 26, 32, 37, 40, 41,
    46, 49–51, 53, 56–58, 72, 74,
    77, 85, 89, 92, 98, 104, 111
Crossmatching, 1, 9

Deoxyhemoglobin, 24, 25, 39, 89
Dextran, 41, 44
Diaspirin crosslinked hemoglobin, 51,
    57, 58, 72, 77
3,4 dihydroxybenzoate, 98
2,3–DPG, 10–13, 19, 21, 23, 24,
    26–28, 31, 37, 38, 74, 75, 79,
    99, 100

Efficacy, 32, 41, 46, 47, 73–74, 77,
    78, 84, 87
Embolism(s), 4, 85
Emergency (-ies), 3, 4, 8, 9, 42
Encapsulated hemoglobin, 13, 14, 26,
    51, 56, 57, 59, 69, 70, 98–99,
    101–102, 104, 107, 110–111
Endothelial cell(s), 68, 69, 70
Endothelin, 77
Endotoxin, 29, 32, 50, 60, 64, 90
Enzon, 19, 20, 31, 72, 80, 111
Erythropoiesis, 5
Erythropoietin, 5, 7, 79
Esophageal spasm, 69
Exchange transfusion, 46, 47

Factor X, 50
First generation blood substitute(s), 1, 3,
    87–88, 90–91, 111–112
Fluosol–DA, 81–83
Fluosol–DA 20, 81–82

Glutaraldehyde, 17, 18, 26, 30, 31,
    33, 35, 38, 74–75

Heme, 23, 25, 31, 92–93
Hemodialysis patient, 77

HemoGen, 84, 111
Hemoglobin, 110
    Binding site for 2,3–DPG, 24
    Heme–heme interaction, 25
Hemoglobin lipid vesicles, 103–106,
    108, 109
Hemoglobin nanocapsule, 13, 63, 65,
    104–109
Hemoglobin outside red blood cell, 12, 31
Hemoglobin–SOD–catalase, 89, 92
Hemorrhagic shock model, 40, 41, 45
Hemosol, 27, 72, 75
Hepatitis, 1, 6, 9, 60, 112
Hill coefficient, 25, 38, 107
HIV, 6, 19, 22, 60, 110–112
HTLV, 6

Immunogenicity, 51
International Society
    for Artificial Cells, 113
International Symposium on Blood Substitute, 113
Intestinal ischemic reperfusion, 95
Intramolecularly crosslinked
    hemoglobin, 16, 18, 32
Intraoperative autologous blood, 7
intraoperative autologous blood, 7
Iron, 5, 91–92, 95, 97

Kidney, 9, 12, 18, 31, 39, 104, 110

Long term storage, 8
Lyophilization, 39
Lysine, 22, 24, 30, 31, 92

Methemoglobin, 30, 31, 32, 39, 73, 76,
    79, 92, 104, 107–109
Methemoglobin reductase system, 107
Microencapsulated hemoglobin, 13, 14,
    35, 40, 56, 85, 99
Modified hemoglobin, 1–3, 5, 8–10, 13,
    19–24, 26, 28, 31–32, 35, 37–39, 44, 47, 49,
    51, 58, 60, 61–65, 67–70
Myocardial infarction, 5

Nitric oxide (NO), 49, 68–70

o–Raffinose human polyhemoglobin, 72,
    75–76
Oncotic pressure, 11, 12, 32, 35, 76, 80
Organ preservation, 4, 46
Orthopedic surgery, 4
Oxygen dissociation curve, 15, 23–25, 38, 99,
    100, 107
    Sigmoidal shape, 25
Oxygen radicals, 90–92, 98
Oxygen–binding site, 23
Oxygent, 84